Writing Effective Ecological Reports

Writing Effective Ecological Reports

A Guide to Principles and Practice

Mike Dean

Pelagic Publishing | www.pelagicpublishing.com

Published by Pelagic Publishing
PO Box 874
Exeter
EX3 9BR
UK

www.pelagicpublishing.com

Writing Effective Ecological Reports:
A Guide to Principles and Practice

ISBN 978-1-78427-241-8 (Pbk)
ISBN 978-1-78427-242-5 (ePub)
ISBN 978-1-78427-243-2 (ePDF)

Copyright © 2021 Mike Dean

The moral rights of the author have been asserted.

All rights reserved. Apart from short excerpts for use in research or for reviews, no part of this document may be printed or reproduced, stored in a retrieval system, or transmitted in any form or by any means, electronic, mechanical, photocopying, recording, now known or hereafter invented or otherwise without prior permission from the publisher.

A CIP record for this book is available from the British Library

Cover image: River Wye, iStock/fotoVoyager.

Contents

List of text boxes viii

Foreword by Mike Oxford ix

Acknowledgements xiii

1. Introduction **1**
Key characteristics 8

2. Competence, qualifications and experience **19**

3. Getting the basics right **27**
General 27
The passive voice 28
Impersonal style 28
Emotive language 30
Accuracy with words describing quantities 31
Forming sentences 32
Forming paragraphs 34
Tenses 35
Certainty of language 37
Punctuation 38
Bullet points and numbered lists 39
Abbreviations/acronyms 40
Species names 40
Brackets 41
Headings and sub-headings 41
Sub-sub-headings 42
Numbering 42
Avoiding double negatives 43
Font size and type, and paragraph or line spacing 43
Highlighting text 44
Headers and footers 44

4. Fact versus opinion 46

5. Report structure 55

6. Making a start 58

7. First impressions and opening lines 65
Title Page or Cover Page 66
Contents Page 68
Introduction 70

8. Getting your facts right 72

9. So what does all this mean? 89

10. Keeping it in proportion 109

11. Tables, figures, photos and appendices 117
Contents Pages 118
Tables 118
Figures, drawings, maps or plans 119
Photos 124
Other illustrations or drawings 127
Appendices 127

12. Creating and using a template 130
Why use a template? 130
What are the drawbacks? 132
Other options 133
Solutions to the pitfalls of using a template 133
How to set up a template 134
How to use a template 136

13. Writing an effective Summary 138

14. PEA or EcIA Reports – what's the difference? 146
PEA Reports 147
EcIA Reports 149
What do these differences mean? 151
When can you submit a PEA Report instead of an EcIA Report with a planning application? 153

15. Writing Environmental Statement chapters **155**
Background 155
Scheme description 156
Assessment methods 156
Structure 158
Cross-referencing 158
Consultation 159
Summaries 160

16. Proofreading, technical review and Quality Assurance **161**

17. Tips for those reviewing reports **173**

18. Referencing sources **179**

19. How long is a report valid? **187**

20. Useful sources of information **191**

Appendix A: Suggested headings and sub-headings for an ecological survey report 194

Appendix B: Suggested headings and sub-headings for Preliminary Ecological Appraisal Reports 196

Appendix C: Suggested headings and sub-headings for Ecological Impact Assessment Reports 198

Appendix D: Suggested headings and sub-headings for Biodiversity Action Plans or Strategies 201

Appendix E: Suggested headings and sub-headings for an ecological monitoring report 202

Appendix F: Suggested headings and sub-headings for an ecological monitoring strategy 204

Appendix G: Suggested headings and sub-headings for ecological method statements 205

Index 209

Text boxes

1. Possible consequences of poor versus good reports 7
2. Fact, evidence, opinion, and professional judgement 10
3. Different types of ecological report, their purposes and target audience 15
4. Competence 20
5. Desk study information – details to be provided 76
6. Typical layout of assessment section of an Environmental Statement chapter versus a stand-alone EcIA Report 159

Foreword

When Mike Dean asked if I would write the foreword to this book, of course I said yes. I wanted to add my enthusiastic endorsement for such valuable new guidance for the ecological profession. However, when I agreed, I thought it would be an easy task. After all, I have been working in ecology and the planning system for over thirty years. I have read and reviewed hundreds of ecological reports and written my fair share too. So, how difficult could it be to pen a short foreword? But as I sit here to write the first words, I've been struck by just how important and timely *Writing Effective Ecological Reports* really is.

All this has taken me by surprise, and my task is a little more daunting than I first imagined. This book deserves the ecological equivalent of a trumpet fanfare. Perhaps a 90-decibel recording of a rampant, roaring bull elephant sent out as an mp3 file to every practising ecologist. It really is that sort of book; it warrants such a proclamation.

Mike has produced a book that should be essential reading for anyone involved in writing or reviewing ecological reports. It should be on everyone's reference shelf. In fact, this book is probably relevant to every single ecologist, as I can't think of one professional role where, at some point, you wouldn't need to know how to write a report *effectively*.

When I read Mike's draft a few months ago, I had both an emotional and instinctive response. First, was envy! Why hadn't I thought of this brilliant idea, as it fills such an obvious gap in the market? That said, it is hard to be envious of someone who I respect and hold in such high esteem. While Mike is the consummate professional, and capable of pin-point clarity over what is right or wrong with a report, he is also extremely humble. And don't take my word for it, just read the opening of Chapter 1. Right from the outset, he avoids preaching from a pedestal and instead concedes that he is capable of making mistakes just like the rest of us. And this is a tone that runs throughout the book. He sets out to offer clear insight into what makes for an effective ecological report from the position of someone who has learned the lessons the hard way.

Setting aside my fleeting feelings of envy, I now come to the deeper instinctive reaction I had when I started to understand what Mike is trying to achieve here. Simply put, he wants us all to be able to write really effective ecological reports. But think about it, that is incredibly ambitious! From my own perspective, gathered over three decades of working in and with local planning authorities all around the UK, I'd say a large proportion of ecological reports fall far short of being fully effective. But why does the need for effective ecological reports stir me at some primal level? I would have to say it's partly to do with self-preservation, along with a burning desire to ensure my great-grandchildren (no, I don't have any yet) are also able to enjoy a future filled with nature in all the diversity I have enjoyed during my life.

It's clear to me that, as ecologists, we need to know how we can communicate, to the very best of our abilities, about the perils facing the natural environment and the means by which, as a society, we can mitigate that damage. And let's be under no illusions, 'us ecologists' have a huge responsibility to help halt the decline of biodiversity. We need to do that here in the UK and anywhere else we may work around the globe. This is one of the main reasons I get out of bed in the morning (although, as I get older, getting to the bathroom is normally the first!). By the way, Mike says he's not keen on authors putting text inside brackets!

At the time of writing this foreword, the year 2020 has been extraordinary for many reasons. I am sure the speed at which the global pandemic has affected just about every aspect of modern life will be a long-lasting mark in the history books. However, I will also remember 2020 as the year we saw signs that humanity was truly waking up to the idea that there is a very real and present danger from the biodiversity and climate emergencies. Scientists everywhere tell us that urgent action is required. And people are beginning to take notice. Politicians are even standing on the world stage and committing their governments to ambitious environmental targets.

We need to turn their rhetoric into action. But that's no easy task. Taking action to avert widespread species extinctions and planet-wide habitat decline is all about communicating the issues and challenges; it's about providing robust assessments of potential ecological damage and harm; it means developing and delivering well-conceived practical solutions that actually halt and reverse declines in biodiversity.

Let's not underestimate the task: halting biodiversity loss at the scale required will mean unprecedented levels of effort, informed by exceptionally rigorous technical and professional capabilities. As ecologists, we have our work cut out.

Here's the thing though. It doesn't matter how good an ecologist we think we are. In fact, it matters not a speck on the kneecap of a daddy-long-legs (that'll be the *Tipulidae* family) whether we are a brilliant entomologist (or any kind of other *-ologist*) if we can't communicate effectively with our audience over why and what specific action is needed.

So, to be blunt, if we do not write effective ecological reports, we will, to varying degrees, have failed in our mission. We will have missed the opportunity presented in that moment to halt further loss of biodiversity. I believe it is Yoda in one of the Star Wars films who says:

> 'Do or do not. There is no try.'

This quote offers a simple lesson in commitment and the power of giving something our all – not just giving it a try! In other words, if we write an ecological report that isn't fully effective, then we are only giving it a *try* and any significant weaknesses in the report may mean:

> 'The protection of biodiversity we do not do!'

I was proud to be part of the team that saw the publication of the first ever British Standard for Biodiversity, this was BS42020 *Biodiversity – A Code of Practice for Planning and Development.* The Standard sets out *what* needs to be done at each stage of the planning and development process to ensure that ecological issues are adequately addressed by those involved. However, what BS42020 does not do is explain *how* ecological reports should actually be structured, formatted and written in order that they achieve the very best outcomes for biodiversity.

This is where Mike Dean has excelled. This book is very much a 'how to do it' manual. He has given careful consideration to what it takes to prepare and write an effective ecological report – one that is truly fit for purpose. He has set out the principles and practices that contribute to a document that should provide clients, decision-makers and consultees with comprehensive and easily digestible ecological information. There is no unnecessary jargon and Mike's writing style is an example that we can all hope to emulate in our own reports; it is always simple and clear.

It is to Mike's credit that he has written a source of guidance that I think everyone will find useful. I believe that ecologists at the start of their careers will find it to be an invaluable guide to steer them along the right path, while also learning how to avoid common pitfalls. Meanwhile, there are little nuggets of wisdom and advice in here that will keep even the most experienced practitioners on their toes. On more than one occasion, I read sections that had me thinking carefully about my own work and whether it measures up to the expected standards.

Finally, in terms of writing quality, this book sets the bar for ecologists. For anyone who genuinely wants to help halt and reverse the loss of biodiversity, this book will show how to write reports that contribute to that noble aim in the most effective means possible.

Dare to do, not try!!

Mike Oxford FCIEEM, CEcol.
October 2020

Acknowledgements

This book is the product of experience gained working as a professional ecologist over the past 23 years. During that time I've been fortunate enough to have worked with a large number of other ecologists and have undoubtedly learnt something valuable from each and every one of them. They have all contributed in some way to the advice that is contained within these pages – normally by pointing out a mistake that I'd made (in a helpful and constructive way), by showing me a better way of doing something, or by simply sharing their ideas. I won't attempt to name them all, for fear of missing someone out.

I do, however, have to mention one person in particular – the late Warren Cresswell. I was lucky enough to get a job working for Warren and his wife Steph in 2001. I wrote a lot of reports over the next few years, and the task of reading and editing them, before they were allowed to be sent to a client, often fell to Warren. He must have had a good stock of red pens because my reports would usually come back covered in annotations. I learnt a huge amount from that experience. I don't recall that I minded having to amend a report that I'd spent ages writing after Warren had covered it in red ink, because I always felt that it was significantly improved by his corrections. And he would take the time to explain why he'd changed something, ensuring that I improved with each report I wrote. I couldn't possibly have written this book without the knowledge I gained from Warren, his eye for detail, and his infectious enthusiasm for all aspects of the natural world. Thank you Warren.

I am also particularly indebted to fellow ecologists and friends Bob Edmonds, James Latham and Mike Oxford, who were kind enough to take the time and trouble to read and comment on a draft of this book. Their suggestions have been invaluable and I am incredibly grateful. I'd also like to thank Paul Chanin for his insightful comments on my text relating to otters.

I would also like to thank CIEEM for allowing me to draw on some of the details in their publications, particularly the suggested headings and sub-headings for ecological method statements (Appendix G).

Chapter 1

Introduction

A bit of background

As is, hopefully, clear from the title of this book, the following pages give advice on how to write 'Ecological Reports'. By this I mean any written report produced on the subject of ecology in a professional capacity. Many will comprise reports used in development planning (such as Ecological Impact Assessment Reports or Preliminary Ecological Appraisal Reports) and, given that this is where I have most experience, these sorts of report were uppermost in my mind when writing this book. However, there are a number of other types of report that professional ecologists write, as I'll describe later. The advice in this book is intended to be applicable to all such reports.

It is perhaps slightly odd to write a book about how to write. From a professional point of view it's also slightly risky, as I'm bound to have made some grammatical errors over the course of trying to explain how not to make mistakes.[1] And there will, of course, be those who disagree with some of the statements I make, or approaches that I suggest. But I know from my experience of writing reports, reviewing reports written by others, writing guidance on report writing, and running training courses on this topic, that many ecologists need a source of advice on this. So that's what I hope this will be.

---

1 You will find a few obviously deliberate mistakes as you read this book. These have been written as examples of what not to do. You may also find some less obvious ones. If you do, let's just say that those are deliberate as well.

There is a difficult balance to strike when writing a book like this. Some of you reading this will have been writing (or reviewing) reports for many years already, and a proportion of what I cover might seem really obvious. I am, after all, trying to provide guidance to those reading this with less experience as well. However, I've tried to make sure that there is plenty of useful advice for the more proficient amongst you, and that each and every chapter contains something that will be relevant to everyone, no matter what their level of competence.

I will come back to the subject of competence regularly throughout this book, with a particular focus on it in Chapter 2. Whilst we're on that subject, you might well be asking yourself, 'Who is Mike Dean, and what makes him competent to tell us how to write a report?' This would be a very fair question, so I'll try to answer it.

I'm an ecological consultant. I've worked as a consultant since 1997, with roles in varying size consultancies from the very small (one or two staff members) to the very large multidisciplinary consultancy, and a few in between. Throughout this period I've been undertaking ecological surveys and producing ecological reports of all different types and sizes. Over the years I've made my fair share of mistakes when it comes to writing reports and been fortunate enough to have others, with far more experience, reviewing and correcting my reports. Since around 2008 I've been undertaking technical reviews of reports produced by others, including more junior members of staff in the same company and more latterly as a subcontracted role to other consultancies.

Since 2011 I've had the opportunity to get involved with writing guidance through the Chartered Institute of Ecology and Environmental Management (CIEEM) including being a member of CIEEM's Professional Standards Committee, tasked with producing guidelines on, amongst other things, ecological report writing.[2] I've been delivering training on report writing for CIEEM since 2015.

None of this makes me necessarily better than anyone else at writing reports but hopefully I've learnt enough about the subject to be in a position to provide some useful advice, particularly in relation to the sorts of report that ecological consultants routinely produce. It's not something that I always

2 CIEEM is a professional body that many ecologists belong to.

find easy even now, and it certainly wasn't when I first started writing reports.

Why is report writing so difficult?

A career in ecology will bring numerous challenges, requiring a variety of skills to be mastered. Professional ecologists need to be competent at a range of skills such as having the ability to identify certain species, or to assess the condition or quality of habitats. They should also have knowledge of relevant legislation, policy, good practice guidelines and the outcomes of recent research, and will often need to have experience of applying and interpreting these in the context of a specific project. And, as if all of that isn't enough, it is essential that they also possess the ability to communicate effectively, both verbally and in writing.

Writing reports, then, is simply another skill that a professional ecologist needs, like learning how to tell that a hole in the ground is a badger sett rather than, well, just another hole in the ground. It does appear, though, that this particular skill is one that many professional ecologists struggle to get to grips with – some by their own admission, whilst others appear to be unaware that their reports are not delivering what they should.

CIEEM occasionally receives complaints made against its members, which can be for a number of different reasons and has published guidelines on ecological report writing to help point its members in the right direction.[3] It also receives complaints made against ecologists who aren't members, but there's obviously much less that it can do about that. One of the main reasons for a complaint against an ecologist, or at least a factor in that complaint, often relates to the quality of the report that was produced.

So why is this? Well, there are a few reasons that are immediately obvious.

Firstly, a written report is the main way that ecologists communicate the outcomes of their studies and assessments. There will often be a significant amount of both field and desk work that's been completed to underpin that

---

3 CIEEM (2017) *Guidelines on Ecological Report Writing, 2nd Edition*. Chartered Institute of Ecology and Environmental Management, Winchester.

report. This may have been done exceptionally well, but the reader won't necessarily realise this – all they see is the quality of the written work.

Secondly, professional ecologists write reports that have an important function for many other members of society, particularly where they are submitted to the Local Planning Authority to inform them of the likely outcomes for biodiversity associated with a proposed development. The Local Planning Authority will be making major decisions based, in part, on an ecologist's report. This means that, in many cases, the report will be subject to a considerable degree of scrutiny by a large audience with wide-ranging viewpoints.

These two reasons combined put a significant amount of pressure on the quality of reports. There's nothing we can do to avoid these. Instead we need to recognise and accept them, doing our best to produce high quality work able to withstand rigorous scrutiny.

Thirdly, there's nowhere to hide with a report. Everyone makes mistakes, but with much of the work done by an ecologist there are often opportunities to correct those mistakes. Perhaps an ecologist completed a particular survey but forgot to record a specific element of the survey work, or neglected to visit one part of the site. That's undoubtedly inconvenient, and likely to cost the business time and money to put right. However, there's a simple remedy – go back to site and fill in the gaps! Of course there can still be a problem if the mistake isn't spotted and rectified soon enough, as some surveys need to take place at particular times of year, but provided the error is quickly identified and resolved there shouldn't be a lasting problem.

The same can apply to a report to a certain extent, provided the author does what they can to maximise the likelihood of any mistakes being spotted and dealt with. However, once a report has been submitted to a client or the Local Planning Authority it can be difficult to correct those mistakes – not impossible, but some of the reputational damage will already have been done. And if nobody spots those mistakes they can result in significant problems for many stakeholders, including the biodiversity resources that professional ecologists should be trying to protect.

Mistakes made in a report can lead to poor outcomes for biodiversity long after the report was written. For example, errors made in a survey report, if not identified and corrected, can find their way into a subsequently written

Ecological Impact Assessment (EcIA) Report. Perhaps the mitigation or 'on-site' protection measures set out in the EcIA Report are incorrect because of the original mistake in the survey report. These measures can become enshrined in planning conditions and the error may never be spotted, or may only be identified once works have commenced and it's more difficult to correct the error.

Fourthly, ecologists, just like any other professionals, have to work within time and budget constraints. As the report is usually the final piece of work the ecologist is contracted to complete, it is often subject to the greatest time pressures. And in some cases, even if sufficient staff time was allowed to complete it at the outset, that 'budget' may have been eaten into by overspends that have had to be accounted for already. This basically means that reports, despite being the most 'externally visible' element of the work, are often written quickly. And when we do things quickly we are likely to make mistakes.

The third and fourth reasons described above are, to some extent, within our control. There are simple things we can do to try to deal with both of these. I'll come back to spotting mistakes in Chapter 16, and to the issue of taking sufficient time in Chapter 6.

The final reason is simply that most ecologists didn't choose this line of work because they like sitting in an office writing reports. The sort of people that pursue a career in ecology are often most at home pulling on a pair of wellies and wading down a river, or crawling around a dark and cramped roof space, or climbing trees, or searching out and sniffing animal poo, or ... well you get the idea! Office work is something to be reserved for cold and wet winter days – it's what we do when we can't do the part of the job we love. This doesn't apply to everyone, of course, but is likely to be a major factor for many.

We obviously can't change who we are, and frankly wouldn't want to – ecologists need to be good at all of those other parts of the job as well. However, there are things we can do to start to overcome this reluctance to spend time in the office. We can take steps to try to make the environment we work in more conducive to writing a report. I'll cover this in detail in Chapter 6. And we can make the whole process of report writing less daunting, partly by doing it more often, and partly by getting better at it. The more often we do something and the more confident we are with it, the less we'll fear it, and the better we'll become. And of course the opposite is equally true – it's a vicious circle that we have to try to break out of.

Ecological reports are used for a variety of different purposes, including informing decisions on whether to authorise a proposed development or not. Such decisions should be made on the basis of appropriate and sufficient information.[4] The information must be robust and communicated in a manner that others, including non-ecologists, can understand. Producing poor or ineffective reports can have serious consequences for biodiversity resources. It can also have negative consequences for those writing the reports, those reliant on them, and the wider ecology profession. On the other hand, good quality reports will have significant benefits, ensuring that decisions are well informed. The possible consequences of poor versus good reports are summarised in Box 1.

A professional ecologist will be asked to write a wide variety of different reports. These will range from a simple report, setting out the results of a survey and the methods used, to a detailed mitigation strategy, a habitat management plan, or a chapter of an Environmental Statement (the report that presents the outcomes of an Environmental Impact Assessment).[5]

This variety of reports gives us yet another problem that we face when writing reports – it's difficult to come up with a standard approach that works for everything. Many companies use a 'standard template' for their reports – this can be a useful starting point in some cases, but comes with its own problems, which are discussed in Chapter 12.

In some cases an ecologist will need to communicate information in the form of a site visit record or a briefing note. These might not be thought of as 'reports' per se but they still require careful consideration, and much of the advice in this book will apply equally to these documents.

---

4 See Section 6.2 'Adequacy of ecological information' in the British Standard *Biodiversity – Code of practice for planning and development BS42020:2013*, published by the British Standards Institution.

5 An assessment under the EIA Regulations. For development projects in England this will normally be the Town and Country Planning (Environmental Impact Assessment) Regulations 2017. Statutory Instrument No. 571. Other (similar) EIA Regulations exist for other sorts of project and in other parts of the UK and Ireland.

Box 1: Possible consequences of poor versus good reports

Who is affected?	Consequences	
	Poor reports	**Good reports**
Author	• Lack of job satisfaction • Limited opportunities for professional development • Increased stress levels	• Job satisfaction • Personal development
Individual or company responsible for producing the report	• Loss of clients • Loss of professional reputation • May be difficult to retain and recruit good staff • Legal action	• Enhanced professional reputation • Improved client relationships • Improved team spirit within a company
Wider profession	• Loss of respect for ecologists amongst developers and planners • Poor public perception of ecologists, and biodiversity resources (such as protected species)	• Ecologists are valued members of a client's team, and are therefore better placed to improve outcomes for biodiversity resources • Clients, planners and the public are reassured that biodiversity resources are being respected
Local Planning Authority (LPA)	• Planning decisions may be ill informed • Approval may be given for poorly designed mitigation measures, or planning applications may be refused due to inadequate information being provided • LPA may fail to secure appropriate measures through conditions	• LPA are able to make well informed decision • LPA can secure appropriate mitigation, compensation and enhancement measures
Client	• Projects may be delayed if ecology constraints are not properly identified • Additional costs can result in budget overspends • Risk of prosecution	• Project risks are reduced
Biodiversity resources	• Damage to, or loss of, resources as a result of decisions based on poor information	• Better protection of biodiversity resources

Key characteristics

Any good professional ecologist has the ability to write an effective report. The first step is to be able to recognise what makes a good, effective report – what are the key characteristics? This isn't about reeling off a list of section headings that should be included (*Introduction, Methods, Results,* etc.) – although that is important and we'll come back to that in a later chapter. No, what I mean by 'key characteristics' is more a series of fundamental principles. These will come up regularly throughout the following chapters, but here's a brief summary to get us started. In my view, a good, effective report must be:

1. **Purposeful** – has clear aims and objectives that meet the expectations of the intended target audience, and that it delivers against;
2. **Targeted** – is written with its target audience in mind;
3. **Well structured** – has a logical flow that makes it easy to follow;
4. **Transparent and truthful** – is open and honest about the data presented, the sources of data, any limitations to collecting the data, any interpretation of the data, and whether a statement is a fact or an opinion (see Box 2 for a definition of these terms);
5. **Robust** – is based on sound data which is sufficient and appropriate to support the purpose;
6. **Justified** – provides suitable evidence for any conclusions reached and recommendations made;
7. **Written by a competent person** – the author is suitably qualified to make the judgements contained within the report;
8. **Impartial** – is not biased towards a point of view that either benefits or disadvantages any stakeholder, including the client;
9. **Proportionate and concise** – provides an appropriately balanced treatment of the issues, with more 'air time' given to those that are more important or more complex, without providing excessive amounts of unnecessary information;
10. **Clear and precise** – sets out the information in a manner that is easy to understand, unambiguous and with attention to detail.

If a report stacks up well against all of these Key Characteristics then it is likely to be an effective report. Of course, judging delivery against each of these is subjective, and we can argue about whether a report is good, excellent, or simply OK. And some of these Key Characteristics represent competing constraints – for example, it can be difficult to be **concise** whilst also being **robust**. There is, then, a balance to be struck. One thing is for certain though, in my mind at least, if a report is found seriously wanting against any of the 10 Key Characteristics then it may not be fit for purpose. And if a report starts to fail against one Key Characteristic then a reviewer might be more likely to questions whether it fails against some of the others as well.

Let's focus on the first two Key Characteristics – being **purposeful** and **targeted**. I've put these at the top of the list, not because they're necessarily more important than the others, but because they are the first things you need to think about when writing a report – before you even put pen to paper (or fingers to keypad). This is really simple, but if you don't do it, the report will almost certainly fail. So, before you start, you must know the answer to the following two questions:

1. What is the purpose of the report?

2. Who is the target audience for the report?

The answers to these two questions have an impact on virtually every decision you'll have to make when writing the report. I have read a lot of ecological reports of different types during my career, and I instinctively know whether the report is likely to be any good or not within a few minutes of starting to read it. If the author hasn't understood the purpose of what they're doing, or who they are writing for, or if they haven't explained that in the first few paragraphs, there isn't a great deal of hope for the rest of the report. How can a report reach an effective conclusion if it isn't clear what it's trying to achieve?

Box 2: Fact, evidence, opinion, and professional judgement

Fact: There are a number of definitions of the word 'fact'. When I use it in this book I'm using it in the sense of *'a truth verifiable from experience or observation'*.[6] In other words it is something known to be true, and which can be proven to be so (or, at least, proven beyond reasonable doubt).

Evidence: I'm going to use the word 'evidence' throughout this book in a very general sense as *'ground for belief or disbelief; data on which to base proof or to establish truth or falsehood'*.[7] It might be, for example: a field survey result; a photograph of a particular feature; a table of data showing dates, times and weather conditions when a survey was undertaken; or an extract from an appropriate reference source which supports an assertion about a species' distribution or behaviour.[8] Personal experience may also be used as evidence, but note that this is less easily verified than the other examples given above and may therefore be considered less reliable.

Opinion: This is defined in the New Collins Concise Dictionary of the English Language as a *'judgment or belief not founded on certainty or proof'*. This definition demonstrates that 'opinion' is clearly distinct from 'fact', as it cannot (or has not) been demonstrated to be true. This is a wide-ranging definition, as it can include opinions that are well supported by available evidence, as well as those that run counter to scientific knowledge and cannot be supported by any credible evidence. When I use the word 'opinion' in this book I mean it in this very general sense – covering both of those scenarios as well as any others in between.

Opinions, by their nature, are personal – as we all know, everyone is entitled to their own opinion! Nevertheless, we are intuitively likely to give more credence to some opinions rather than others. This will depend on:

---

6 New Collins Concise Dictionary of the English Language (London and Glasgow: Collins, 1982).

7 Ibid.

8 See Chapter 4 for a more detailed discussion of what constitutes an 'appropriate reference source'.

1. The extent to which an opinion is supported by available evidence;
2. The reliability of that evidence;
3. The extent to which an opinion is shared by peers; and
4. The level of knowledge and experience of the person expressing the opinion in relation to the subject matter.

Some opinions will be widely held by those with a reasonable knowledge and experience of the subject matter, and have a good deal of supporting evidence. At the other end of the spectrum there will be opinions that are not widely held and don't have supporting evidence or are perhaps even contrary to the evidence. I'll look at this issue in more detail in Chapter 4.

Professional judgement: This is defined in the British Standard *Biodiversity – Code of practice for planning and development BS42020:2013*, as *'use of accumulated knowledge and experience in order to make an informed decision that is clearly capable of being substantiated with supporting evidence'*.

As professional ecologists we are expected to show *'sound professional judgement'*. In other words, we should be expressing opinions which can be supported by evidence and which are widely shared amongst our peers, highlighting situations where there is any evidence to the contrary or a widely held alternative viewpoint. We should also only be making judgements on issues that we have sufficient knowledge and expertise of, particularly where there is a lack of published evidence.

For an in-depth discussion of the term 'professional judgement' and its use, I will refer you to an article published in CIEEM's *In Practice* in March 2016 entitled 'Pragmatism, Proportionality and Professional Judgement'.[9]

---

9 CIEEM Professional Standards Committee (2016) Pragmatism, Proportionality and Professional Judgement. *In Practice* 91: 57–60.

Different ecological reports have different purposes and different target audiences. Box 3 sets out some of the most common ones. Many professional ecological reports will have a role to play in protecting biodiversity resources in relation to a proposed development. Different reports might be used at different stages, such as:

1. Helping a developer to determine the feasibility of a project, or to design it with minimal impact on biodiversity (a Preliminary Ecological Appraisal (PEA) Report,[10] for example);

2. Allowing the Local Planning Authority to make an informed decision on whether or not to consent a proposed development, given the outcomes for biodiversity that are likely to occur as a result (an Ecological Impact Assessment (EcIA) Report,[11] for example);

3. Providing sufficient detail to enable the protection of biodiversity during construction of the proposed development, based on information that may have been unknown at the stage that the planning application was submitted, such as a detailed design or a construction programme (an Ecological Mitigation Strategy or Ecology section of a Construction Environmental Management Plan (CEMP), for example).

Understanding where a particular report fits into an overall process like this, knowing what it's going to be used for, and who is going to use it, are fundamentally important to writing a good report. I've met many professional ecologists over the course of my career who still don't have a good understanding of this, despite being relatively experienced and having spent years writing the sorts of report that I've referred to above. It is, unfortunately, a recipe for a poor report.

Once you've got used to identifying the purpose and target audience each time you write a report, you'll find that the process will become much easier. We'll return to this again in Chapter 6.

---

10 See CIEEM (2017) *Guidelines for Preliminary Ecological Appraisal, 2nd Edition*. Chartered Institute of Ecology and Environmental Management, Winchester.

11 See CIEEM (2018) *Guidelines for Ecological Impact Assessment in the UK and Ireland: Terrestrial, Freshwater, Coastal and Marine*. Chartered Institute of Ecology and Environmental Management, Winchester. Version 1.1 updated September 2019.

Moving on to the third of those Key Characteristics – being **well structured**. Try thinking of a report as a journey on which you need to take a reader (specifically, any member of your target audience). The journey starts with a purpose – so set out the aims of your report in the *Introduction*. And the journey ends (in most cases) with a *Conclusion* that must relate back to the purpose you started with. The rest of the report is a series of logical steps taking you from the initial purpose(s) to the conclusion(s). There are exceptions to this, of course, but it holds true in most cases.

You need to think about the reader. Will they understand the steps you're taking? Will they challenge them, and want to see the justification for an assertion that you've made or an opinion that you've expressed? The steps need to be logical and flow in an order that makes them easy to follow. And the steps need to be relevant to the purpose and take a reasonably direct line towards the conclusion. This isn't a novel where you can take a reader off at a complete tangent just for interest's sake or to keep them guessing about 'Who Dunnit?' Anything that doesn't fit into those logical steps doesn't really belong in the report. Of course, you may want the report to fulfil a secondary purpose, but if that's the case you need to be clear about it.

Key Characteristics 4, 5 and 6 on the list are being **transparent and truthful**, **robust** and **justified**. Many ecological reports will contain both facts and opinions (see Box 2 for a definition of these terms). There are some exceptions, such as a simple report setting out the results of a survey and the methods used – this may only contain facts, such as what was done, how it was done, who did it, when it was done, and what the outcomes were. However, even a report like this will often contain a *Discussion* or *Recommendations* section. And most other reports that ecologists write will certainly contain such sections (or their equivalent). So, it's absolutely vital that a report:

- Is clear about whether a statement is a fact or an opinion;
- Is honest when it comes to the facts it contains and any interpretation of those facts; and
- Contains appropriate data, or other supporting evidence, to justify any opinions or assessments.

Of course, the opinions expressed in a report will be based on the author's experience, and so a report written by someone with insufficient knowledge or expertise will be, at best, unreliable. At worst it could be harmful. This makes it essential that a report is **written by a competent person** (Key Characteristic 7). We'll come back to the other Key Characteristics later on, but we're going to focus on the issue of competence in more detail in the next chapter.

Box 3: Different types of ecological report, their purposes and target audience[12]

Title of report	Purpose	Target audience
Ecological survey report	• To present the methods and results of a specific survey	• Client • Design team • Other ecologists who may use the report to inform their work, such as an EcIA Report
Preliminary Ecological Appraisal (PEA) Report	• To identify ecological constraints to a particular development scenario as well as opportunities for delivering enhancement or biodiversity benefits • To identify the mitigation measures and licences likely to be required • To identify any further surveys needed	• Client (developer) • Design team
Ecological Impact Assessment (EcIA) Report or Ecology/Biodiversity Chapter of an Environmental Statement	• To present an assessment of the likely (significant) effects of a development proposal on ecological features/biodiversity • To allow a determining authority to ascertain whether the proposal accords with relevant planning policy and legislation • To allow a determining authority to write planning conditions or obligations (where necessary) to secure mitigation, compensation and enhancement measures	• Determining authority (normally the Local Planning Authority) • Nature conservation consultees, such as the relevant Statutory Nature Conservation Body, local Wildlife Trust, etc. • Other stakeholders (such as the client) • Members of the public with an interest in the site

// ctd.

12 Based on the following:

CIEEM (2017) *Guide to Ecological Surveys and Their Purpose*. Chartered Institute of Ecology and Environmental Management, Winchester.

CIEEM (2017) *Guidelines on Ecological Report Writing, 2nd Edition*. Chartered Institute of Ecology and Environmental Management, Winchester.

Title of report	Purpose	Target audience
Ecological Appraisal	There is no widely agreed definition of the purpose of such a report. It might be an EcIA Report, or a PEA Report (see above), or something in between the two. It is therefore generally best avoided as a report title – if used it needs to be made clear what its intended purpose is, and who it is written for.	
Biodiversity Action Plan/Strategy	• To set out the aims and principal approaches to conserving and enhancing biodiversity within a specific geographical area (e.g. for a County or for a development site)	• Client • Determining authority (where written to discharge a planning condition) • Nature conservation consultees • Those responsible for delivering management actions
Research Report	• To report on the outcomes of a specific piece of research	• Client • Professional ecologists or other interested parties
Guidance or advice document	• To provide advice or guidance on a particular issue, such as an assessment process, habitat management, or survey or mitigation techniques for a protected species	• Wide ranging, including professional ecologists and amateur naturalists
Ecological Monitoring Strategy	• To set out the aims and specific methods for ecological monitoring, normally in relation to post-construction monitoring at a development site	• Client • Determining authority (where written to discharge a planning condition) • Nature conservation consultees • Those responsible for undertaking the monitoring
Ecological Monitoring Report	• To report on the results of ecological monitoring, normally post-construction monitoring at a development site	• Determining authority (or licensing authority) • Nature conservation consultees • Client
Habitat Management Plan or Landscape and Ecological Management Plan	• To set out the aims of management of a site, or the retained and new habitats within a development • To detail the management prescriptions necessary to achieve the stated aims	• Client • Determining authority (where written to discharge a planning condition) • Nature conservation consultees • Those responsible for delivering management actions

Title of report	Purpose	Target audience
Ecological Constraints Report	• See PEA Report	• See PEA Report
Ecological Mitigation Strategy or Ecology section of a Construction Environmental Management Plan (CEMP)	• To set out details of ecological protection measures to be implemented as part of a construction project • May also include specific details of ecological mitigation measures (see also 'Method Statement' below)	• Determining authority (where written to discharge a planning condition) • Client • Contractor • Site ecologist/Ecological Clerk of Works
Method Statement • To accompany a protected species licence application or • For non-licensable works (sometimes referred to as a Precautionary Working Method Statement)	• To set out details of site works required, and any restrictions on site works, normally in relation to a protected species • Normally includes a detailed programme of works, as well as details of machinery, equipment and personnel required	• Statutory Nature Conservation Body (where the Method Statement has been written to accompany a protected species licence application) • Contractor • Site ecologist responsible for overseeing works • Client
Reports used in the Habitats Regulations Assessment[13] process There are a variety of different terms used and reports can be produced to present the outcomes of a screening process or to inform an 'Appropriate Assessment'	• To inform an assessment of the effects of a project on a European Site or European Offshore Marine Site either alone or in combination with other plans or projects	• Competent authority (as defined in the Regulations, such as the Local Planning Authority) • Relevant Statutory Nature Conservation Body • Other nature conservation consultees, such as the local Wildlife Trust, etc. • Other stakeholders (such as the client) • Members of the public with an interest in the site or project

// ctd.

13 In England and Wales, the stages of assessment which must be undertaken in accordance with the Conservation of Habitats and Species Regulations 2017 (as amended) and the Conservation of Offshore Marine Habitats and Species Regulations 2017 (as amended) before deciding whether to undertake, permit or authorise a plan or project that may affect a European Site or European Offshore Marine Site. The term is used in other parts of the UK and Ireland although the specific legislation referred to will differ.

Title of report	Purpose	Target audience
Ecology Proof of Evidence	• To express the personal opinion of an expert witness on ecological issues at an Inquiry	• Planning Inspector (or equivalent) • All other interested parties at an Inquiry
Biodiversity (Net) Gain Report, or similar The specific report titles to be used at different stages of the assessment process (such as feasibility stage or design stage) had not been determined at the time of writing	• To provide the outcomes of an assessment of predicted biodiversity losses and gains associated with a proposed development using an approved metric • To provide the data that underpin the calculations, and a commentary on that data where necessary • To demonstrate that the development and any biodiversity gains have been designed in accordance with published standards[14] or guiding principles on Biodiversity (Net) Gain[15]	• Determining authority (normally the Local Planning Authority) • Nature conservation consultees, such as the relevant Statutory Nature Conservation Body, local Wildlife Trust, etc. • Other stakeholders (such as the client or relevant landowners) • Members of the public with an interest in the site
Site visit record[16]	• To record the key outcomes of a site visit	• Varied
Briefing note, or equivalent[17]	• Varied	• Varied

14 At the time of writing, a British Standard on this topic has been drafted and is the subject of a consultation process.

15 CIEEM, CIRIA, IEMA (2016) *Biodiversity Net Gain: Good practice principles for development.* CIEEM, CIRIA, IEMA.

16 I've included 'Site visit record' and 'Briefing note, or equivalent' in this list to reflect the fact that there are instances where ecologists communicate in a written form that may not routinely be considered a 'report'. It's difficult to provide details of the purpose and target audience for these as they will vary in different cases, but that's not to say that the author of such a document doesn't need to consider these – they do! And the advice in this book will, to a large extent, be equally applicable to such documents.

17 See note 16 above.

Chapter 2

Competence, qualifications and experience

Assessing competence

It's always a thorny issue, but we can't sweep it under the carpet: how do we decide if someone is sufficiently **competent** to write a professional ecological report? As we'll see, it's not an easy question to answer.

One of the key problems with determining competence to write a professional ecological report is that there are essentially two separate 'skills' or 'competencies' that are involved:

1. Presenting information in a written form; and
2. Having sufficient knowledge and experience of the subject matter to be able to undertake surveys and make informed judgements (and, of course, we could subdivide this into multiple separate skills or competencies – such as one for each species or habitat).

These are difficult to disentangle when we read a report, but both are vital. An ecologist might have a significant amount of expertise at dealing with a particular species for example, but may struggle to present the information and their assessment in a readable way. Another ecologist might be good at crafting a report to make it 'user-friendly' and appear professional, but we might question whether they have sufficient expertise to have made the judgements contained with it.

Having sufficient knowledge and experience of the species and habitats covered by the report is vital, but providing advice on judging that is beyond the scope of this book. In this chapter, then, I'm going to try to focus on the first of those skills or competencies I listed above – that of presenting information in a written form (although I acknowledge that it's difficult to separate this from the other skill or competency completely when assessing the competence of a report author). There are no specific qualifications or assessment criteria to determine competence when it comes to report writing. We can start with CIEEM's definition of competence (see Box 4). This suggests that, in essence, you're **competent** to write reports when the reports you write are consistently good.

Of course it can be difficult to assess your own competence. How do you know that your own reports are good? Or recognise weaknesses in your own work?

To an extent this is a matter of opinion, but whose opinion? In the first instance it may be your manager or a more senior member of staff. But in actual fact they aren't the final arbiter – that role falls to the main audience for whom the report was written, so it will vary for different reports (see Box 3 in Chapter 1).

Box 4: Competence

CIEEM define someone as being competent at a particular skill if they:

- Know what to do;
- Know how to do it;
- Know when to do it;
- Know why you do it;
- Can do it consistently well; and
- Know when to seek help and advice.

(from: https://cieem.net/wp-content/uploads/2019/02/Competency-Framework-web-FINAL.pdf)

In many cases, then, one might assume that the decision on whether a report is good or not will rest with the person who commissioned it (the client). If they think the report does what it is supposed to do, then it's an effective report, and the author is clearly **competent**. So, if you don't get bad feedback from your client then you've passed – haven't you? Well, not quite. There are a number of problems with this as an approach to judging whether you've attained a sufficient level of competence:

1. A client might not actually read an ecology report that they've commissioned. Ouch! That's a pretty shocking statement and a little bruising to our egos. It's sadly true though. Many clients will read an ecology report but many won't. This may be because they don't have specialist ecological knowledge, so will judge a report on the basis of whether it is 'successful' or not – does the Local Planning Authority consider the report to be acceptable, for example?

2. A client may be unhappy with an ecology report even though it's a good competent one – it may tell them things that they don't particularly want to hear.

3. If a client is not used to reading ecology reports then they might find it difficult to differentiate between a good one and a bad one.

4. A client may be unhappy with the service they receive from an ecological consultant for reasons that don't relate to the quality of the report, but they're unlikely to take the time to fill out a detailed questionnaire to allow you to tease that out.

5. If a client is unhappy with a report they receive from an ecologist they may simply find themselves another ecologist – you may never know that they thought your report was terrible!

6. The client isn't the main target audience for all ecology reports.

Having said all of that, one of the best judges of the quality of an ecology report is a well informed client. Some developers simply want an ecology report that doesn't raise any issues (even if there are issues that should be raised). However, in my experience these are becoming rarer. The majority of developers want an ecology report that highlights the issues in a language that they can understand – plain English.

The other 'best judge' is another well informed and experienced professional ecologist. One who is used to writing or reviewing the sort of report you're producing. In the case of an ecology report written to inform a planning decision, this may well be the Local Planning Authority's ecologist (although not all local authorities have one, and not all of them have extensive experience of planning casework). If they think that the reports written by a certain individual regularly meet the 'standard' they are looking for, then it's reasonable to assume that the author is fairly **competent**.

And another ecological consultant may have a need, at some stage, to read and provide feedback on one of your reports. Dependent on how experienced they are this might give you a good steer on the quality of your work. Of course, if they have an axe to grind, then they might look to pick holes in it. But then again, if they have an axe to grind and they struggle to find any serious faults, then you're probably getting it pretty much right.

The best way to determine your own competence then is to test the quality of your reports. What do other professional ecologists think of them? Ask an ecologist doing similar work to you to read them and comment. What do well informed clients think of them? This can be a bruising experience, particularly if you thought your reports were really good, and someone then finds some significant issues with them – you have to be prepared to hear criticism. And if we're being honest it's surely impossible to write a report that is completely fault-free, so they're bound to find something wrong – the question is whether those faults are fundamental (does it fail to deliver against the Key Characteristics discussed in Chapter 1?) or are they less significant?

One of the principal reports that a professional ecologist is likely to have to write is an Ecological Impact Assessment (EcIA) Report. For projects that fall under the auspices of the Environmental Impact Assessment (EIA) Regulations, this will be the 'Ecology' or 'Biodiversity' chapter of an Environmental Statement (or equivalent). The EIA Regulations include a requirement that such reports are '*prepared by competent experts*', and that they are accompanied by a statement '*outlining the relevant expertise or qualifications of such experts*'.[1] Incidentally, there's also a similar requirement in the EIA

1 In England: The Town and Country Planning (Environmental Impact Assessment) Regulations 2017. Statutory Instrument No. 571. Paragraph 18(5). Similar requirements are included in the relevant Regulations in other parts of the UK and Ireland.

Regulations for the assessors of that report, who need to have '*sufficient expertise*' to examine it.[2]

So why then, is there no qualification of competence for writing or reviewing 'Ecology' or 'Biodiversity' chapters of Environmental Statements, or their equivalents? How else can we be certain that the document has been produced by '*competent experts*', or reviewed by someone with '*sufficient expertise*'.

In time it's possible (perhaps even likely) that such a qualification will be developed. And that may well be extended to include Ecological Impact Assessment Reports for development projects that don't fall under the auspices of the EIA Regulations. After all, although it's not a statutory requirement in such cases, the principles of ensuring that those writing and reviewing such reports are 'competent' or 'have sufficient expertise' would still be considered good practice. However, until such a qualification is developed we will have to make do with judging our own, or each other's, competence. Qualifications for producing the other ecological reports listed in Box 3 in Chapter 1 may also be useful but without legislation or a policy requiring it, these are relatively unlikely to be developed – at least not in the near future.

Becoming competent

So how do we achieve this hallowed status of competence? There isn't really a shortcut – it takes time, and will come with experience.

I doubt any professional ecologist was producing high quality reports when they started their career – I know I didn't! Most of us will have graduated from a degree course where we learnt to write in an 'academic' style. Writing reports in the style that ecological consultancies typically produce can take some getting used to.

So, given that you've probably already got a good understanding from your education and current or previous employment, by the time you've finished

---

2 In England: The Town and Country Planning (Environmental Impact Assessment) Regulations 2017. Statutory Instrument No. 571. Paragraph 4(5). Similar requirements are included in the relevant Regulations in other parts of the UK and Ireland.

reading this book you should have many of the traits of being **competent** at writing ecological reports following the definition in Box 4:

- You should know what is entailed in writing a report;
- You should know, in general terms, how to write a report;
- You should know when to write a report (and the specific types of report to write in any given situation); and
- You should know why you write reports – understanding what they are going to be used for.

This leaves us with:

- Being able to write reports consistently well; and
- Knowing your limits and when to seek advice.

Hopefully this book will help with the first of these two aspects. And the more chances you get to write reports the better.

Knowing your limits is a general character trait – some people are naturally more cautious in this regard than others. What is important, though, is that you have a source of advice when you need it – we're all more likely to ask for advice if we think we'll receive something helpful. This book will provide you with some of that advice, but can't cover every difficult situation you face.

It obviously helps to gain experience of producing reports in a company or organisation that provides plenty of support for its less experienced ecologists by giving its more senior members of staff a mentoring role. The most effective way of learning how to do anything is to try doing it, get it wrong (or partly wrong), and be shown how to do it correctly. This is certainly the way that I learnt how to write professional reports – discuss it with that 'mentor', have a go at drafting it and submit it to them for their review, receive a version back covered in red pen highlighting all the horrible mistakes you've made, correct it, and (hopefully) learn from those mistakes.

Importantly with report writing, it is vital that errors are explained thoroughly. It won't always be immediately obvious why a particular sentence in a report is wrong – it might be grammatically and factually correct, but might be 'wrong' for quite subtle reasons, such as because it doesn't give the correct emphasis to the reader. Indeed, the same sentence might have been 'correct' in one report, but considered 'incorrect' in another. This can be difficult for the less experienced to understand, and can be quite demoralising. It therefore needs to be carefully explained. And if you are in the position of receiving a draft report with lots of amendments scribbled on it, and don't understand where you went wrong, then ask the person who undertook the review and suggested the changes.

There's an important point to be made here about having a positive 'mindset'. It's the same as anything really – if you start from a perspective of thinking that you will never write good reports then you're unlikely to improve significantly. Be proactive when it comes to seeking help and advice. If you're unsure about how to tackle a report, or about why something you wrote in a previous report was amended – ask.

You will have to push yourself to improve. Provided you have an appropriate level of support and guidance from someone who is sufficiently **competent**, and are given the opportunities to have a go, you'll make rapid strides. For those individuals who work in companies without very experienced individuals, or where those individuals can't or won't give sufficient support, the process of improvement is likely to be much slower. And errors may actually be reinforced where the individual in the 'mentor' role hasn't got to grips with their own mistakes when it comes to writing reports.

I regularly hear from ecologists that attend training courses I run that their boss keeps 'correcting' their reports in a way that is incorrect as far as they are concerned (and in some cases contrary to the good practices set out in guidance documents or that I cover during the course). This is particularly worrying, as it means that the chances are very high that there are professional ecologists working in the industry, with relatively little experience, being led astray on the subject of report writing by insufficiently experienced, but more senior, staff members.

When do you reach this level of competence then? Because of the range of different types of report that an ecologist will have to write in their career, you're likely to become **competent** at writing certain types of report before you become **competent** at writing others. The reports you have to produce more often are likely to be better than ones you've never had to write before. But once you become more experienced you'll be able to take on any type of ecology report, as you will routinely apply the same guiding principles to each (such as those Key Characteristics that I discussed in Chapter 1).

So, ecological consultancies should treat the skill of report writing in the same way that they might treat a field skill, such as undertaking a bat survey. The report needs to be produced by someone who is already **competent** but, just like the bat survey, they can be assisted by someone who is still learning this particular skill. The more experienced 'mentor' can allow the less experienced person an increasing degree of autonomy with each report they write, as they demonstrate that they're improving. But the mentor is still there to sign off the final document, and to offer advice and support during its initial drafting.

How do small companies without suitably experienced staff, or individuals who work for themselves, acquire this skill? Well, again, it's like learning any new skill – you can attend training courses (but make sure they're run by experienced individuals), you can seek out good examples of reports and review them to see how you could improve your own reports, or you can ask a more experienced ecologist from another company to mentor you, by guiding you on how to approach specific reports and perhaps reviewing them for you.

Any approach to developing this skill is likely to cost in terms of both time and money – you have to be willing to invest! It is, after all, one of the most important skills that a professional ecologist needs, particularly if they work as an ecological consultant.

It could reasonably be expected by an ecological consultancy that, when they employ a new member of staff who has a degree, several 'A' levels and a string of GCSEs, that individual will have a good grasp of the English language – they will be able to spell, have a good understanding of grammar and be able to use punctuation properly. Sadly, it appears, this isn't necessarily the case. So in the next chapter we're going to look at some of the basic writing skills in the context of ecological reports.

Chapter 3

Getting the basics right

General

In order to write a good report you need to possess a good grasp of grammar, be able to spell the words you use and be able to punctuate properly. Clearly, a few spelling or punctuation errors are not the end of the world. But if a report is littered with such mistakes then it will be difficult to read – the reader will keep having to stop and re-read the sentence, and this will interrupt the flow. It will be, a bit dis-jointing witch will make it; annoying and theres noting worse that that! A bit like that last sentence. These sorts of error are largely beyond the scope of this book, although in Chapter 16 I will deal with a few of those that occur regularly in ecology reports.

I'll start off by looking at some of the basic style conventions that we need to follow when writing reports – as opposed to writing a letter or a novel.

I'll then tackle some of the basic style and structure issues that are specific to the reports that professional ecologists write. Most ecologists will have a scientific background, and will therefore be used to writing scientific reports or papers. Whilst there are similarities between the requirements of a scientific report and a 'professional' ecological report, there are also numerous differences.

The passive voice

Reports should be written using the 'passive voice' rather than the 'active voice'. This means that we write sentences focusing on the thing that experiences an action, rather than the thing that performs the action. This sounds quite complicated but it is something most of us are probably familiar with already. An example of a sentence in the active voice would be:

> *Bob Smith surveyed a pond for great crested newts.*

Putting this into the passive voice, we might write:

> *A pond was surveyed for great crested newts.*

The focus of the sentence is on the pond that has been surveyed, rather than Bob Smith who undertook the survey. It's not that we aren't interested in who undertook the survey – It's just the style of how we traditionally write reports. As it is we are actually interested in who did the survey, so in reality we might write:

> *A pond was surveyed for great crested newts by Bob Smith.*

Or:

> *A pond was surveyed for great crested newts. The survey was undertaken by Bob Smith.*

Impersonal style

In most cases, the reports we produce should be written in an impersonal style. For example, even if I carried out the surveys being reported, I wouldn't write:

> *I carried out the water vole and otter surveys.* (Personal style and the active voice)

Or:

> *The water vole and otter surveys were carried out by me.* (Personal style and the passive voice)

Instead I'd write:

> *The water vole and otter surveys were carried out by Mike Dean.*

This is probably a fairly obvious example. So what about when we express an opinion or make a recommendation in a report? These must be personal to the author of the report – so wouldn't it make more sense to write the relevant sentences in a personal rather than impersonal style? Well, maybe, but we don't. It's simply a convention – it's the accepted way of doing it. Presumably this is because it tends to sound more professional. For instance, we wouldn't write:

> *I recommend that bat boxes are installed as a mitigation measure.*

Or:

> *I consider it likely that dormice are absent from the site.*

Instead we'd write:

> *The installation of bat boxes is recommended as a mitigation measure.*

Or:

> *It is considered likely that dormice are absent from the site.*

This rule holds true for most ecological reports, but there is an exception – Proofs of Evidence. These are documents that set out an ecologist's 'professional' opinion and are normally submitted as written 'evidence' to inform a planning inquiry. For this reason they are often written in a personal style. This can sometimes be mixed with the more conventional approach of writing in an impersonal style and in the passive voice.

Emotive language

Remember those Key Characteristics we discussed in Chapter 1? The eighth on the list is **impartiality**. Ecological reports then need to be careful not to sway the reader to a particular point of view – they need to be unbiased. This can be difficult, particularly given that we have to provide an interpretation of the basic facts in most reports (and therefore have to express an opinion). We'll look at this in more detail in the next chapter, but the key point here is that we use a fairly plain form of the English language to express those opinions, rather than using lots of unnecessary, flowery, pointless or, perhaps, inflammatory adjectives.

Did the words 'unnecessary', 'flowery', 'pointless' or 'inflammatory' make that last sentence more interesting to read, than if I'd simply written 'rather than using lots of adjectives'? Well, yes, they probably did. But could they also have led you into thinking that adjectives are worthless or perhaps even slightly subversive? This is an example of the use of emotive language; professional ecological reports need to avoid this wherever possible.

I'm not saying that you can't use any adjectives – just make sure that the ones you use are either objective, such as an adjective describing colour, or are supplemented with facts. For example:

> *There was a tall, unmanaged hedgerow located in the centre of the site.*

Whether a hedgerow is tall or not is subjective, and all hedgerows will have been managed at some point in the past, so the term 'unmanaged' is imprecise. You could also argue that the 'centre' of the site might mean something different to different people. This could therefore have been written as:

> *There was a tall (7–8m high) hedgerow, which did not appear to have been trimmed for several years, located in the centre of the site (indicated by H1 on Figure 2).*

There are some words that we will use in a report that could be construed as being emotive. Wildlife legislation, for instance, uses words like 'recklessly', 'intentionally' and 'deliberately'. So we might use these words when we write out the legal protection afforded to a given species. However, we need to be careful to use such words only in that specific context.

Accuracy with words describing quantities

There will be examples in any report where the author needs to express scale in relation to size, distance, time or frequency. Always think carefully before using a vague term, and replace it with a specific and accurate term instead wherever possible. For example:

> *Bat surveys of the site were undertaken regularly during the period April to October 2018.*

The term 'regularly' suggests that the surveys took place at evenly distributed intervals – that may, or may not, have been the case. But what the reader wants to know is how many bat surveys were undertaken and at what intervals. It would therefore have been more informative and accurate to write:

> *Bat surveys of the site were undertaken on seven occasions, at approximately monthly intervals, commencing in April 2018, with a final survey visit in October 2018.*

In reality the reader (or at least some of the readers) will want to know the specific dates when the surveys were undertaken, which may be provided in an appendix, for instance. I'll discuss this issue further in Chapter 8.

Here are some other examples of vague words or phrases used to describe quantities:

- Lots, loads, considerable, many, some, few, several, a large number of – try to replace these with an actual number.
- Often, frequently, multiple times, on a number of occasions – again, try to replace these with a number.
- In the surrounding area, nearby, close to the site, in relatively close proximity – replace with a distance from the site, or a defined geographical area. There will also be circumstances where the direction from the site is as relevant as the distance; in such cases it should be stated (north, north-east, east, south-east, etc.). A drawing or figure showing proximity to the site will also be useful in this context (see Chapter 11 for a discussion on the use of drawing or figures).

There will, of course, be occasions when you have to use vague words or phrases, such as where you simply can't give a specific number. When you do so, make it clear that the word is being used to deliberately indicate a vague concept. For example, you could start a sentence with 'In general terms, ...' – but check first to make sure it is acceptable to be non-specific in that particular case.

Forming sentences

There are general grammatical rules for forming sentences that I won't go into here – it's taken as a given that those rules will be followed. Instead, I'm going to focus on the general principles that relate to professional reports.

Firstly, sentences should be short. The longer a sentence is, with multiple bits of punctuation, indicating separate clauses, and sub-clauses, the less likely anyone is to understand the key points you're making and, importantly, perhaps even vitally, the greater the chance of them giving up before they get to the end. If a sentence needs to be read several times to ensure you understand it then it probably needs breaking down into more than one sentence – a bit like the one I've just made you read.

Secondly, try to make the information in a sentence flow in a manner that is logical to the reader. Having short sentences should help this. For example:

> *Badgers are a material consideration in determining a planning application as they are a legally protected species, which has setts present within the site boundaries, that could be affected by the proposed works.*

This sentence includes a number of separate pieces of information that are not delivered to the reader in a logical manner. The first piece of information to express is that there are badger setts present on site, so this fact should come at the start of the sentence. The next thing to inform the reader of is that those setts could be affected by the proposed works, so that should come next. Thirdly we need to tell the reader that badgers are legally protected, and finally arrive at the point that, as a result of this legal protection, they are a material consideration in determining a planning application. The sentence could have been more clearly written as two shorter (and reordered) sentences, such as:

There are badger setts within the site boundaries that could be affected by the proposed works. Badgers are a legally protected species, making them a material consideration in determining a planning application.

Thirdly, you should avoid starting sentences with joining words such as 'although', 'but', 'and', 'or'. These words should be used within a sentence rather than at the start of it, as they refer directly to the previous thing that's been written. In some cases it can be tempting to start a sentence with 'although', and the sentence can be made grammatically correct. The problem is that this is normally achieved by turning the sentence back-to-front. This tends to result in the ideas not flowing in a logical manner.

You need to be particularly careful when using words like 'although' and 'however' – make sure that they are actually needed. For instance:

Although skylarks were recorded nesting outside of the site boundary during the survey, it is considered likely that they are absent from the site as the habitats present within the site boundaries are not suitable for use by skylarks.

The problem here stems from what the author is implying by the word 'although' in this context, which might not be the same as what the reader thought they were implying. For example, the reader's logic might say to them – if skylarks were recorded nesting outside the site, that implies that they weren't recorded nesting in the site, so therefore they are likely to be absent, so why do you need to say 'although'?

This is making the reader work hard to understand the point being made. It would be clearer to break the sentence down into two separate sentences and take the reader through the logical thought process we want them to follow. For example:

Skylarks were recorded nesting outside of the site boundary during the survey. They were not recorded nesting within the site and the habitats present within the site boundaries are not suitable for use by skylarks. It is therefore considered likely that they are absent from the site.

As you will see from this example, there's often no need to use the word 'although' at all.

Forming paragraphs

Paragraphs also need to be simple. A paragraph should contain a single idea, so that all the sentences within a paragraph are covering the same concept or part of the discussion.

A paragraph could contain a single sentence (like this one).

When a paragraph exceeds 6–8 lines it starts to become ineffective. The reason for this is simply that our mind will wander when reading a long paragraph. We might lose our place and have to start again from the beginning. So when a paragraph starts to become longer than 6–8 lines it's a good idea to try to break it down into separate paragraphs.

It can also be helpful to break a paragraph down into multiple shorter paragraphs when a sentence in the paragraph relates to one of the other sentences, but not to all of them. For example:

> *A desk study was undertaken by contacting the Devon Biodiversity Records Centre to request existing ecological information for the site and its surroundings (up to 2km from the site). This included information relating to non-statutory designated nature conservation sites and records of protected species or species of conservation concern. Bat records within 5km of the site were also requested from the local Bat Group. In addition, information relating to statutory designated nature conservation sites within 10km of the site was obtained from Natural England's 'MAGIC' website. The District Council's Biodiversity Officer was consulted (by telephone) in relation to the site on 3 May 2018. The data search was carried out during April 2018.*

It would be clearer to separate out the sentence relating to consultation with the District Council's Biodiversity Officer into its own paragraph, as it deals with something different from the rest of the paragraph (i.e. consultation rather than a data search). It would also be helpful to split the statements relating to different data sources into separate paragraphs.

Paragraphs also need to be capable of standing alone without requiring the reader to have read the previous one. So, in general, they shouldn't start with words such as 'however', 'although' or 'therefore'. If you find yourself doing this, you probably need to combine the paragraph you're writing with the previous one. If that makes it too long, then try phrasing the information in a different way.

Tenses

Different reports, and different sections of the same report, will need to be written in different tenses. The main rule is to avoid mixing tenses as far as possible, although this will be difficult in some cases.

As a general rule, most of the information set out in the *Introduction* should be written in the present tense. For example:

The aim of this report is to ...

The site is located ...

The proposed development comprises ...

There are likely to be some exceptions to this. For instance, you might write '*this report has been produced by ...*'. This will obviously have to be expressed in the past tense.

Methods and *Results* sections should be written in the past tense. For example:

A Phase 1 habitat survey was undertaken ...

The presence of great crested newts was confirmed ...

A badger sett was recorded ...

The use of the past tense rather than the present tense is important in *Results* sections, as it highlights to the reader that the information being described was correct at a particular moment in time (in this case, at the time that a survey was undertaken) but specifically avoids giving the impression that the information is automatically still correct. There was a badger sett there when you did the survey, but that doesn't guarantee that there's still one there now.

It is acceptable to deviate from the past tense in the *Methods* and *Results* sections where the scenario described above doesn't apply and where it sounds odd to write in the past tense. This is normally the case where we need to refer to the status of something that is unlikely to have changed since the data were collected. For example:

The site was located at Ordnance Survey Grid Reference ...

This sounds odd, as the site is presumably still in the same place, so it would make more sense to write '*The site is located ...*'.

The woodland within the site boundaries was designated as a local wildlife site ...

This could be construed as suggesting that the woodland is no longer designated as a local wildlife site. Assuming that this isn't the case, it would probably be clearer to write '*The woodland within the site boundaries is designated as a local wildlife site ...*'.

Ecological Impact Assessment (EcIA) Reports don't normally include a *Results* section, but instead use the results of ecological surveys to present an assessment of baseline conditions – this is a description of what is likely to be there at the time of any impacts arising from a development, and is therefore normally written in the present or future tenses. In some cases it can be appropriate to include statements written in the past tense – where the author wants to stress that the predicted baseline conditions are reliant on a survey result from the past, which is subject to change in the future. However, the use of the past tense in *Baseline Conditions* sections should be minimised.

EcIA Reports will also include a prediction and assessment of the likely impacts of a development and proposals for required mitigation or compensation measures. The sections relating to these parts of the report should be written in the future tense.

Most other sections of a report will normally be written in the present tense. For example:

The population of slow-worms using the site is considered likely to be of county importance.

It is recommended that 10 bat boxes are installed on retained trees within the site.

Policy B7 of the Local Plan is of relevance to the proposed project.

Certainty of language

Professional ecological reports used to inform a decision maker on whether or not to consent a proposed development project will need to provide 'certainty' that certain proposed actions will be undertaken.[1] For example, an Ecological Impact Assessment (EcIA) Report, describing the mitigation measures to be implemented in relation to a particular biodiversity resource in the event that planning permission is granted, should use positive language. For example:

The following mitigation measures will be implemented ...

The detailed tree planting plan will need to include ...

The contractor responsible for demolition of the buildings will need to produce ...

Phrases that don't provide certainty will generally need to be avoided in EcIA Reports, such as:

The developer could consider ...

It may be appropriate ...

Protective fencing might be installed ...

In some professional ecological reports there will be inherent uncertainties, in which case words such as 'may' and 'could' are more likely to be appropriate. For example, a Preliminary Ecological Appraisal Report may need to provide some guidance on likely ecological constraints, but further surveys may be required before a firm set of mitigation measures can be determined.

When writing a report you will therefore need to think carefully about the level of certainty needed by that type of report and the level of certainty possible in the circumstances, using the appropriate words in each case.

---

1 See Section 6.6 'Providing certainty and clarity for the decision-maker and the applicant' in the British Standard *Biodiversity – Code of practice for planning and development BS42020:2013*, published by the British Standards Institution.

Punctuation

Reports are meant to be simple to read. It shouldn't be too difficult to understand the key messages. Short sentences, with minimal punctuation, are therefore better than long ones. In general then, try to write sentences that need little punctuation.

Semicolons seem to be particularly difficult to get to grips with. They have two main uses.

The first use is to separate items in a list, which should start with a colon, such as:

> *The following tree species were recorded within the hedgerow: hazel; ash; hawthorn; blackthorn; and elder.*

You could separate the species list into bullet points, and the punctuation would still work (see 'Bullet points and numbered lists' below). In a bullet point list it would be equally acceptable to simply do away with the semicolons completely.

The second use is more complicated, and often results in errors. It is to link two related statements by putting them into a single sentence. For example:

> *Water voles are listed as a species of particular importance for conservation in the Nottinghamshire Local Biodiversity Action Plan; they are rapidly declining in the UK as well as across the county.*

The use of the semicolon emphasises the link between the decline of water voles and their inclusion in the Local Biodiversity Action Plan. The rule for this use of a semicolon is that you must be able to replace it with a full stop and the resulting two sentences would still be grammatically correct. So why not just have a full stop? Well, to be honest, in many cases that would be better. In general I'd try to avoid using a semicolon in this way, going for the full stop option instead. Only use this sentence structure on occasions where you specifically want to stress the connection between two statements. Try not to use it repeatedly within the same report – partly because it makes it more difficult to read the sentence and partly because if you keep using this sentence structure the reader will become 'immune' to it, and won't detect the stressed connection between two clauses in circumstances when you want them to.

Bullet points and numbered lists

Bullet points are an extremely good way of presenting lists in a report. They're easier to read than a list presented within a sentence, as per the example used above under the heading of 'Punctuation'. However, they lose their impact if there are too many bullet points within a list (10 or more is probably too many) or if there is too much text associated with each one – if you're writing a full sentence or more then the information should probably be presented in a different way.

Numbered lists can be used instead of bullet points, particularly where you want to highlight the importance of reading the list in order, or that there are a specific number of items in the list. For instance:

> *The layout of the proposed development has been designed to avoid impacts on two key ecological features:*
> 1. *The area of ancient woodland in the eastern half of the site; and*
> 2. *Pond A, which is used as a breeding site by great crested newts.*

However, if the order that you read the listed items in doesn't matter, or the list is not a finite one, then I'd stick with bullet points. For example:

> *The following tree species were recorded within the hedgerow:*
> - *hazel;*
> - *ash;*
> - *hawthorn;*
> - *blackthorn; and*
> - *elder.*

In the example above, does it matter whether each bullet starts with a capital letter or not? Some styles in your word processing software will suggest a capital letter for you. And do you need the semicolons at the end of each bullet? In my view, neither really matters – provided that you're consistent. The style of bullet point you use is also entirely up to you.

I have a couple of other personal rules for using bullet points. Firstly, I try to avoid having sub-bullet points – a series of bullet points under another bullet point, such as:

> *The following ecological features will be considered further in this assessment:*
> - *woodland*
> - *hedgerows*
> - *reptiles*
>   - *slow-worms*
>   - *grass snakes*
> - *bats*
>   - *common pipistrelles*
>   - *soprano pipistrelles*
>   - *brown long-eared bats*

Not only does this look messy, it also lacks precision – what about reptiles other than grass snakes and slow-worms, and bats other than those listed?

My other personal rule is to avoid having a bullet point list that extends across a page, unless absolutely necessary.

Abbreviations/acronyms

These can be used in reports provided that the full text is provided at first mention, with the abbreviation/acronym provided in brackets afterwards. The abbreviation can then be used for all future mentions. For example:

> *The Royal Society for the Protection of Birds (RSPB)...*

> *The Design Manual for Roads and Bridges (DMRB)...*

Species names

Common names of species should be used in reports (where they exist – some species only have scientific or Latin names). Capital letters will need to be used where the common name contains proper nouns, such as American mink, Himalayan balsam, or Daubenton's bat. Some authors will capitalise all common names irrespective of proper nouns (e.g. Common

Frog) whereas others don't (e.g. common frog). There is no strict convention, so either approach is fine provided that it is adopted consistently.

It is normal practice to provide the scientific or Latin name of a species in italics after the first use of the common name. This is to avoid confusion where a species has more than one common name. Latin names can be put in brackets, or not – again, there's no specific convention, so either is fine as long as it's done consistently.

Brackets

With the exception of providing abbreviations/acronyms or Latin names, I try to avoid putting text in brackets in my reports. It implies that a statement is not of specific relevance and is only being included as an afterthought. I would certainly avoid having an entire sentence in brackets.

As a general rule then, avoid sentences with lots of text in brackets (if it's important text, why put it in brackets), particularly multiple sections in brackets (which is difficult to read) (and definitely don't have really long sections in brackets that are longer than the actual sentence they refer to and clearly pointless).

Basically, don't do what I did in that last paragraph.

Headings and sub-headings

Most reports will include section headings (*Introduction, Methods, Results,* etc.). Each section may be further divided into subsections using sub-headings. For example, *Methods* may be divided into *Desk study methods, Field survey methods,* and *Assessment methods,* or similar.

As a general rule I would normally number both section headings and sub-headings, such as:

2. Methods
2.1 Desk study methods
2.2 Field survey methods
2.3 Assessment methods

These will help guide the reader through the report and should be logically ordered.

Headings, sub-headings, and sub-sub-headings for that matter (see below) should appear on the same page as the first paragraph that follows them.

Sub-sub-headings

In some cases it will be helpful to break the text under a sub-heading down into sub-subsections. For example, under the sub-heading of '2.2 Field survey methods' I might have sub-sub-headings for 'Extended Phase 1 habitat survey', 'Great crested newt survey', 'Breeding bird survey' and 'Reptile survey'. Alternatively, I could do away with the sub-heading of '2.2 Field survey methods' and have '2.2 Extended Phase 1 habitat survey', '2.3 Great crested newt survey', etc.

I tend not to number sub-sub-headings as this means that paragraph numbers, if you use them, will be quite long – unnecessarily long in the context of most reports (see below).

Numbering

Other than numbering section headings and sub-headings then, what else needs to be numbered in a report?

Well, in my view you should number every paragraph. I realise that I've just written that in a paragraph that isn't numbered, but this is a book rather than a report. There will be certain paragraphs in a report that need to be referred to – in a covering email to the client, in a telephone conversation with the local authority's ecologist, or under cross-examination at a Public Inquiry. Paragraph numbers will make this significantly easier.

Fairly obviously, every page of a report should be numbered. For ease of reference it makes sense to assign a nominal page number even to pages where the number won't be displayed, such as where a figure is included. Whether the page number appears at the top of the page or the bottom, centred or to one side, and the style of it, is entirely up to you.

Every table, figure and photograph should also have its own number. Exactly how you do this is also up to you, but it should be done logically and consistently. The two widely used options are:

1. To simply number them sequentially: Table 1, Table 2, Table 3, etc.

2. To number them according to the section of the report they appear in – The two tables in Section 1 would be numbered Table 1.1 and Table 1.2; the only table in Section 3 would be numbered 3.1, etc.

Avoiding double negatives

You should avoid using more than one negative word, such as no, not or a word prefixed by 'un-' in a single sentence. This is because it makes a sentence difficult to understand, and open to misinterpretation. For example:

It is unlikely that the site provides unsuitable habitat for slow-worms.

Does this mean that the site is likely to provide suitable habitat for slow-worms? Or something different?

Font size and type, and paragraph or line spacing

The type of font you use in a report is entirely up to you. No specific rules on this one. To an extent, so is font size. The only caveat on font size is not to go too small. Many readers will read a report on a screen, where they can adjust the size of the document to ensure the writing is large enough for them to read it, irrespective of how good their eyesight is. However, some will still want to print a report and read from a hard copy. The font size therefore needs to be sufficiently large to allow the average person to read it. I'd suggest font size 11 or 12 in most cases.

You need to allow a reasonable size space between paragraphs. The lines within the paragraph also need to be spaced to a certain extent. There are no rules about what this spacing should be – provided that paragraphs are sufficiently spaced to ensure they are clearly distinct.

Highlighting text

Headings, sub-headings and even sub-sub-headings should be highlighted in some way. You can underline, put them in bold, or put them in italics, or a combination. It's up to you. The only rule is to adopt the same type of highlighting consistently through the document for a given level of heading/ sub-heading.

It's also a good idea to highlight direct quotations or extracts from other documents in some way to ensure clarity over the source of the text.

> *Putting a block of text in italics works well for this. It can also be useful to align the text slightly differently from the rest of the text to distinguish it – a bit like this paragraph.*

This might apply to extracts from legislation or planning policy documents for example.

Personally, I try to avoid highlighting individual words or statements in the main text of a report by, for instance, putting them in **bold** or *italics* or <u>underlining</u> to add stress. That shouldn't be necessary in the majority of reports, although the sharp-eyed amongst you will notice one example I give later in this book where I do exactly that (but it is the exception rather than the rule). You will have noticed that I have used highlighting throughout this book – in most cases not to add stress but to distinguish the context of certain words. For example, I've put section headings of reports in italics (as well as capitalising them), so that you can distinguish instances of normal usage of words like summary, introduction or conclusions from instances where they are being used in the context of a section heading. I've also put those Key Characteristics of reports in bold for the same reason.

Headers and footers

You can design the headers and footers on each report page as you see fit. These can be helpful in circumstances where a single page is printed out of a report – they give it context to allow anyone sent that page to work out where it came from. Standard details like the title of the report, date or version number, and company/organisation that produced it, can be useful for this reason (along with the page number – see 'Numbering' earlier in this chapter).

There is obviously more that could be said about the basics of writing reports, but where do you stop? Well, in my view, round about here. So, now that we've got the basics (or at least some of them) out of the way we can move on to something more 'technical'. In the next chapter I want to expand on the discussion I started in Box 2 (Chapter 1) about the difference between facts and opinions, as this is fundamentally important for any ecological report.

Chapter 4

Fact versus opinion

The vast majority of professional ecological reports will contain a mixture of facts and opinions (see Box 2 in Chapter 1 for a definition of these terms).

Some reports will contain relatively few statements of opinion, such as a basic survey report. This sort of report might set out the details of the survey methods undertaken and the results of those surveys – essentially, the facts. However, the same report might also contain a discussion of the implications of the survey findings, or recommendations to the client, or a statement on the adequacy of the survey techniques used, or an assessment of the survey's limitations. When we make such statements we are inevitably expressing our opinions.

Other reports will contain rather more in terms of opinion, such as an Ecological Impact Assessment report, which will include an assessment of the baseline conditions, possibly including predictions of the status of a species at some point in the future, the likely impacts of a proposed development project and whether these will affect the conservation status of an ecological feature (such as a habitat, species population or designated site), an assessment of the likely success of proposed mitigation measures, and conclusions on whether the project accords with relevant planning policy and nature conservation legislation or not. Such statements are clearly all going to be based on the author's opinions.

So, it's accepted that reports will contain both facts and opinions. And in many cases this will be a requirement. Importantly, though, a report must be written in a manner that clearly distinguishes between the two, and doesn't try to pass opinion off as fact (or vice versa). And the opinions expressed should be supported by evidence, or be *'clearly capable of being substantiated with supporting evidence'* (see the definition of professional judgement in Box 2 (Chapter 1)).

Thinking back to those Key Characteristics we covered in Chapter 1, the ones of specific relevance here are being **transparent and truthful**, **robust**, **justified**, **impartial** and **clear and precise**.

Let's start with the facts. In the context of a professional ecological report these might include:

- Basic details about a site, including its location and boundaries;
- Basic details about a proposed project, including the client's details, and the specific activities proposed;
- A description of the methods used for any desk study or field surveys, including dates, study areas, details of personnel involved, and (where relevant) times and weather conditions for each survey visit;
- The list of possible offences (and defences) in relation to a given protected species;
- Data collected through desk study or field survey.

Facts such as these should be presented as **clear**, **precise**, unambiguous statements. In order to be **transparent**, **truthful** and **impartial** we need to make sure that we don't hide any facts that might be relevant. This sounds straightforward enough in theory. However, in order to be effective a report also needs to be **concise** and **targeted**, which would seem to suggest that irrelevant facts should be excluded. The problem then is that the report author needs to determine what's relevant and what isn't, and this means that even the facts that are presented can be based on an opinion. In addition, the author will need to summarise some information. This can also be problematic. As soon as you summarise a set of facts you are having an influence on the reader's interpretation.

Basic details about a site or a proposed project

This sort of information is, hopefully, fairly easy to deal with. You should provide the relevant information in a suitable form.

The location of the site should be provided in the most widely accepted and used format – in the UK this normally means an Ordnance Survey Grid Reference. For accuracy this should be a six figure Grid Reference (with the two letter code included also), such as SU149853. In some cases a ten figure Grid Reference may be given. Normally the Grid Reference of the site's approximate centre (or centroid) is provided.

Site boundaries and location should also be provided on a plan, using a suitable base map or aerial photograph, appropriately labelled and at a helpful scale, so that the reader can interpret the information.

Methods

Providing an accurate description of desk study and field survey methods is also something that most ecologists will be used to doing, and is covered in more detail in Chapter 8. This section of a report is largely factual and therefore relatively uncontroversial normally. However, there are a few key areas where opinions come into the reporting of survey methods, such as:

1. Survey design and its accordance with good practice guidelines
   Undertaking surveys that are in accordance with good practice guidelines is obviously important. However, guidelines are open to interpretation and accordance with guidelines can therefore be a matter of opinion. For this reason it's vital that statements such as '*these surveys were undertaken in accordance with the relevant good practice guidelines*' are qualified. What are the specific guidelines being referred to? Which specific parts of the guidelines have been followed? And, where there are options for different levels of survey effort for example, a justification of why a particular option has been selected.

2. Extent of study area
   The search area for desk study records, or how far a survey must extend beyond a site's boundaries, is open to debate. Often you will see statements such as '*a 1km search area was considered appropriate*'. Statements

like this need to be qualified – sound reasons for the search area selected should be provided. Otherwise this is simply an unqualified opinion, and could open up a report to criticism, as data have been collected on the basis of that opinion.

3. Suitability of weather conditions and timings
   The time of day when a survey was undertaken, the date it took place, and the weather conditions at the time, are facts. Whether or not these were suitable in the context of the survey being undertaken is an opinion. So, stating that a survey '*took place during suitable weather conditions and at an appropriate time of day*' needs to be supported with evidence. The specific dates, times and weather conditions should always be provided for any survey where these might affect the outcomes (dates will be relevant in all cases, times and weather conditions will be particularly relevant for some surveys but not for others). If the dates, times and weather conditions accord with relevant good practice guidance documents then that should be stated. The unqualified opinion of '*the reptile survey was undertaken during suitable weather conditions and at an appropriate time of day*' can thus be turned into a simple statement of facts:

   > *The dates, times and weather conditions for each of the reptile survey visits are provided in Table 4. These are consistent with the recommended periods of the year, times of day and weather conditions for undertaking reptile surveys as set out in the relevant good practice guidelines (Froglife 1999).*[1]

4. Surveyor experience or competence
   I raised the issue of competence in Chapter 2 – specifically in relation to competence of the author of a report. Some ecological reports will make an attempt to describe the competence of the author (as they should). Many ecological reports will make reference to the competence of the surveyors that collected the field data, or the competence of the ecologist that should undertake a recommended action (a further survey, or the oversight of an element of mitigation, for example). In terms of survey methods then, this relates to the competence of a surveyor.

---

1 Froglife (1999) *Reptile survey: an introduction to planning, conducting and interpreting surveys for snake and lizard conservation. Froglife Advice Sheet 10*. Froglife, Halesworth.

As with competence for writing a report, it can be difficult to define competence for undertaking a specific survey. What makes a surveyor competent? Well, to a large degree, it's a matter of opinion. That's unavoidable. So, terms like '*a suitably competent ecologist*' or '*an ecologist with appropriate experience*' are problematic, as this is an unqualified opinion. When a surveyor's competence is described the description should make reference to evidence, such as their level of experience, whether they hold a relevant licence or not, or whether they've attended training courses. This then becomes a series of facts, and any reader of the report can make their own judgement on the likely competence of the individual concerned.

Legislation

I've read many ecological reports where the author has effectively summarised the legal protection applied to a particular species. I can understand the desire to edit the specific wording of the legislation to make it more succinct and less turgid to read. However, in my view this is a mistake. The specific wording of the legislation is important. Whether the term '*breeding or resting site*' or '*place of shelter or protection*' is used, for example, could be important in determining whether an offence has been committed. Determining what constitutes each of these is also something that's open to legal argument. And whether an act needs to be undertaken '*deliberately*', '*intentionally*' or '*recklessly*' to be an offence is also important – each of these words will have a specific meaning in legal terms.

For these reasons I take the view that you either simply state that a given species is protected under a particular piece of legislation (the Wildlife and Countryside Act 1981 (as amended) for example), or you write out the applicable legal offences in full. Anything in between these two options risks misinterpretation.

Results

Clearly, the vast majority of ecological reports need to be underpinned by data – any information relevant to the report needs to be included. We'll look in detail at appropriate ways to present desk study and field survey results in Chapter 8.

Deciding on the data to present and the data to leave out of a report is perhaps the most difficult decision. How do you encourage the reader to focus on the information of most relevance, without deciding for them what they might consider to be relevant. The most effective way to negotiate this tricky balancing act is to provide all of the survey data in an appendix to the report, only giving a brief mention of the items considered to be irrelevant in the main text, with reference to their inclusion in the appendix.

For instance, field voles and common shrews may have been recorded during surveys for reptiles. These could be noted in a table of survey results in a column headed 'non-target species recorded' (or similar) provided in an appendix to the report. They may then be referred to briefly in the *Results* section with a statement that they are not considered to be relevant (assuming that's the case).

And general habitat surveys might identify habitats of no real consequence to the assessment being undertaken. However, it is still likely to be appropriate to describe them for completeness. Again, the information could be moved to an appendix, and then cross-referred to in the main text.

Expressing opinions

There will be lots of occasions in any report where an opinion is required. There are two key things to remember in relation to this:

1. Opinions expressed in reports should be supported by the 'facts' presented in the report or by appropriate reference sources or, ideally, both.

2. The wording of the report should make it clear that a given statement is an opinion and what it is supported by.

For example:

> *Otter footprints were recorded at three locations on the River Avon within the site boundaries (see Figure 1), confirming that this species uses the site.* [This sentence contains facts and a supposition that is supported by the facts.]

> *No great crested newts were recorded during the survey. The level of survey effort and the techniques used follow current good practice guidelines for confirming the presence or likely absence of this species, based on Chapter 5 of Natural England's 2001 Great Crested Newt Mitigation Guidelines. The likely absence of great crested newts is therefore assumed.* [The first and second sentences are facts that can be corroborated by comparing the level of survey effort with the guidelines referred to. The third sentence is clearly an opinion (given the use of the word 'assumed') that is supported by the first two sentences.]

We need to avoid making statements that are completely unsupported. We also need to avoid making statements that imply that an opinion is supported when actually it isn't. For example:

> *No field signs of hedgehogs were recorded during the extended Phase 1 habitat survey. They are considered likely to be absent from the site.*

The first sentence is a fact. The second sentence is an opinion. The two sentences are written in a way that implies that the opinion is supported by the fact. However, given the difficulty in finding hedgehog field signs, an absence of field signs recorded from a simple walkover survey is unlikely to be a reliable way of determining the likely absence of the species, and there is no published guidance (as yet) that would support this opinion. This is therefore incorrect and potentially misleading. If appropriate reference sources and data were used this might have been rewritten as:

> *No field signs of hedgehogs were recorded during the survey. However, searching for droppings or footprints of hedgehogs is not a reliable means of confirming the likely absence of hedgehogs (Cresswell et al., 2012).[2] There are desk study records of hedgehogs within 1km of the site (data provided by the Local Environmental Records Centre) and the site provides suitable habitat for this species. They are therefore assumed to be present within the site boundaries on a precautionary basis.*

What then, constitutes an appropriate reference source? This is a surprisingly difficult question to answer.

---

2 Cresswell, W.J., Birks, J.D.S., Dean, M., Pacheco, M., Trewhella, W.J., Wells, D. and Wray, S. (2012) *UK BAP Mammals: Interim Guidance for Survey Methodologies, Impact Assessment and Mitigation*. The Mammal Society, Southampton.

We could, of course, list a whole series of published 'good practice guidelines', and for some species these probably are the appropriate reference sources. However, not all habitats, species, or groups of species, have such published guidance. In some cases there are also several pieces of guidance relating to a single species or species group with conflicting advice. And to make matters worse, this is also a moving target – new guidelines may be published, or published research may suggest that one part of a guidance document is no longer relevant, whilst the remainder is. CIEEM has published a list on its website of what it considers to be the most up-to-date and appropriate guidance documents relating to survey, mitigation, management and monitoring for a range of different habitats and species. The intention is that this will be regularly updated. It provides a useful starting point but clearly can't be a completely exhaustive list.

It's also worth making the point here that good practice guidance documents themselves need to be clear about the difference between facts and opinions, and the evidence base for any recommendations they contain. It is for this reason that I included them in Box 3 in Chapter 1 – although they will have a different structure to most ecology reports, the general principles discussed in this book apply equally. CIEEM has produced a set of principles for producing good practice guidance,[3] some of which overlap with the Key Characteristics listed in Chapter 1.

There may also be published research papers in peer-reviewed journals that should be considered as an appropriate reference source, where they are directly relevant. These will be particularly relevant in the following situations:

1. Where the research paper relates to a species or habitat that doesn't have a good practice guidance document;

2. Where the research paper has been published subsequent to the most up-to-date guidance document, and it therefore provides relevant information that was not available at the time that the guidance document was written (particularly if the subject matter relates to an area identified as one requiring further research); and

---

3 CIEEM (2016) *Principles of Preparing Good Guidance for Ecologists and Environmental Managers*. Chartered Institute of Ecology and Environmental Management, Winchester.

3. Where the research paper relates to a previously unknown or unstudied aspect of ecology, survey technique, ecological impact, mitigation measure or management activity.

Beyond this there may be published articles in periodicals that are not peer-reviewed. It will be important to be careful about the use of such articles, as they may simply provide an opinion with limited scientific evidence to back it up. Nevertheless, they can still be useful and relevant in some cases.

There may be occasions when an opinion needs to be provided without any (or very limited) supporting facts or reference sources – it is particularly important that this situation is made **clear** to the reader. For example:

> *The proposed culverting of a section of the stream may have a fragmentation effect on the local water vole population, as individuals within the two parts of the population either side of the culvert could be less likely to move through the culvert regularly. However, there is no relevant published research allowing a proper assessment of this impact, and the current good practice guidelines for water voles state that this is an area requiring further research (Dean et al., 2016).[4] In this case, therefore, it has been assumed that ....*

We'll look at this issue again in later chapters, but for now let's move on to discuss the overall structure of ecological reports.

---

4 Dean, M., Strachan, R., Gow, D. and Andrews, R. (2016) *The Water Vole Mitigation Handbook (Mammal Society Mitigation Guidance Series)*, edited by Fiona Matthews and Paul Chanin. The Mammal Society, London.

Chapter 5

Report structure

The first step in writing a report, after having determined your purpose and target audience (see Chapter 1), is to set out the structure. What are the headings and sub-headings that would be appropriate?

For some types of report there is some guidance on this – CIEEM's *Guidelines on Ecological Report Writing* include a suggested structure for Preliminary Ecological Appraisal Reports and Ecological Impact Assessment Reports.[1] However, these suggested structures will need to be adapted to suit individual circumstances.

Some of the other reports listed in Box 3 (Chapter 1) will also have a defined structure. Method Statements to accompany protected species licence applications, for example, will often need to follow a structure defined by the relevant licensing authority. And Ecology chapters of Environmental Statements will need to follow the same structure as other chapters of the same document – the structure will therefore often be dictated by an individual responsible for co-ordinating the overall document (see Chapter 15).

The important thing to consider when determining the structure of a report is to ensure that it flows logically. For professional ecological reports this traditionally means something like this:

---

1 CIEEM (2017) *Guidelines on Ecological Report Writing, 2nd Edition*. Chartered Institute of Ecology and Environmental Management, Winchester.

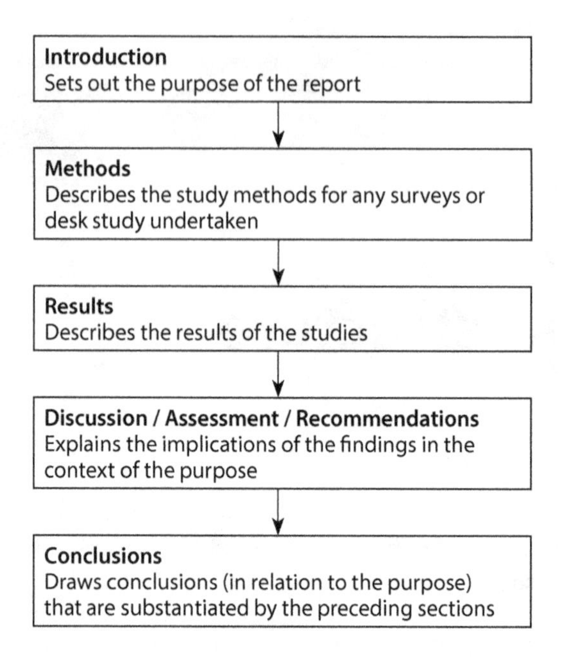

These section headings might be appropriate for a basic ecological survey report or a Preliminary Ecological Appraisal Report, but won't work in all cases. The initial headings of *Introduction, Methods* and *Results* are fairly standard for such reports. The difficulty comes in determining how to deal with the next section or sections, which might include some form of assessment or evaluation, a discussion of the results, or a series of recommendations.

In some cases it can help the flow of the report if you combine *Results* and *Discussion* or *Evaluation*. It's helpful, though, to separate out the *Recommendations* into their own section (assuming that the purpose of the report requires recommendations to be made – they won't always be required) or combine them with the *Conclusions* section. Otherwise recommendations can become lost amongst the rest of the discussion.

There will, of course, be other section headings as well. Firstly, the majority of reports will need a *Summary*. This should precede the *Introduction*, so that it is the first part of the report (following the *Title Page* and *Contents Page*). And there will normally be a *References* section and possibly also a series of appendices.

Ecological Impact Assessment (EcIA) Reports will need to have different headings, but the general flow of the report will be the same. EcIA Reports will normally have a section headed *Baseline Conditions* following the *Methods* section rather than *Results* and *Discussion*. These sorts of report will also need to identify potential impacts, describe the proposed mitigation, and draw conclusions on the significance of any effects. They are also likely to need separate sections on *Biodiversity enhancement, Biodiversity benefit* or potentially *Biodiversity net gain* (dependent upon where the project is located and the requirements that apply) and possibly also *Compensation* and *Cumulative effects*.

There are also reports that will need to follow a completely different structure, such as management plans, method statements or good practice guidance documents, for example.

Deciding on sub-headings can be more difficult. However, getting the sub-headings right can be really important in making sure that the report flows logically. This is where a degree of flexibility is needed. The same series of sub-headings won't necessarily be appropriate in every report of the same type.

Suggested headings and sub-headings for a range of different ecological reports (but not all) are set out in the appendices.

Chapter 6

Making a start

The office environment

So, what is the ideal location for writing a report? Well, personally, I'd say that there are three vital factors to consider:

1. It should be somewhere with immediate access to the resources that you need;

2. There should be sufficient space to spread out maps, survey results, reference sources, etc.; and

3. It should be relatively quiet and undisturbed.

The issue of having access to resources is an important one, and will limit where you can feasibly write a report. Writing reports away from the office – perhaps at home, or in a delightful hotel in a motorway services somewhere en route to a bat survey – is tempting, as it is likely to be quieter than the average office. There may also be space to spread out. However, this only really works if you are able to accumulate all of the resources you're likely to need and take them with you, which will not be possible in many cases.

It can be difficult to predict the specific resources you're likely to need when you first start writing a report. You'll almost certainly need access to the Internet, as well as the file containing survey results, plans, and information from the client. But which particular reference sources will you require? Many of these are now available online, which certainly helps, but not all are. And having a hard copy to refer to is much easier in many cases.

You could, of course, write the report without some of these reference sources, leaving a note to yourself to complete certain paragraphs when you're next in the office. This works fine if you only have to skip ahead in one or two places, and where the 'skipped' information is a matter of detail of limited consequence to the conclusions of the report or the logical steps leading to those conclusions. It doesn't work well if:

- You're missing key reference sources when you write the initial draft, and have to skip significant sections of the methods, limitations, and recommendations sections (or equivalent) as a result; or

- Are missing some important information about the project or key field results and therefore can't tackle the sections that follow.

Trying to write a report in these circumstances will lead to mistakes, as you'll have to write it in multiple 'sittings'. It is often also a frustrating experience.

In most cases, then, you'll need to be in the office to ensure you have access to all the relevant reference sources to allow you to write an effective report. And you should ensure that you have a reasonably large, clear and uncluttered desk (note to self!) to allow you to spread out all of the necessary resources.

The problem with the office in many cases, though, is that it can be anything but 'quiet and undisturbed'. Large open plan offices can be great for discussing the implications of a particular survey result, or planning the next survey visit, but having to listen to everyone else's conversation can be very distracting. Especially for the easily distracted!

The larger the office, and the more people in it, the more of a problem this will be. So for those ecologists working in small offices this might be less of an issue. Those working in larger offices may need to take some specific steps to try to improve the situation. Here are some ideas:

- Wear headphones and listen to some 'background' music;

- Select a day for writing a report when you know most people will be out of the office;

- Be flexible with your working hours (most ecologists have to be anyway for undertaking field surveys) so that you are in the office outside of normal working hours, when things should be quieter – you'll obviously need to clear this with your manager and agree how you will 'balance' your hours, so that you don't just end up working very long days;

- Keep a quiet area of the office as a 'hot desk' that can be used by anyone, when they need to be away from distractions to write a report;

- Politely ask everyone else in the office if they can keep noise, phone calls, etc. to a minimum for a specific period (you obviously don't want to have to do this too often).

Of course, there will be some distractions that are of your own making rather than anyone else's, which you will also need to deal with. It's a good idea to close down your email and turn off you phone. Beyond that it really just comes down to willpower. If you want to be distracted you'll find something to distract you:

- A normally mundane event happening just outside the window where you sit will become suddenly fascinating;

- You'll decide that it's your turn to make tea or coffee for everyone in the office; and

- Despite previously having shown a complete disregard for all company imposed filing systems, you've decided that the project file is not sufficiently well arranged and go off in search of a hole puncher, dividers and as many of those clear plastic wallet things as you can lay your hands on.

This is all part of the fear factor – being able to make rapid strides through the first few sections of the report will certainly help overcome some of this. Making sure you start with the right sections (see 'Parts of the report to tackle first') and that you've got key bits of information at your fingertips (see 'Key things to know before you start') will reduce the likelihood of your mind searching for one of those distractions. Before we get onto those, there are a few other things to mention about the office environment.

Firstly, think about your seating position. You'll need to be able to sit comfortably at a desk for a long period.

Secondly, screen size is also important. You're likely to need to look at reference documents or plans or photographs at the same time as writing your report, so having a screen large enough to accommodate this (or multiple screens) will be vital. You don't want to have to keep minimising the report to view another document.

Finally, before we move away from the office environment, you need to make sure that you allow sufficient time and that you factor in regular breaks. You might not write an entire report at a single 'sitting', but you should make significant progress each time you sit down to write. Writing a few paragraphs before dashing out of the office to do something is unlikely to result in a good report. Taking breaks from the report are also vital – these might be a break to make a coffee, a quick walk outside to clear your head, or even just a break from writing the report by switching to some other item of work. If I have a report to write I'll start on it first thing in the morning then, after a couple of hours, I might spend half an hour reviewing my emails and dealing with some quick tasks, before returning to the report.

Parts of the report to tackle first

There's a strong temptation when writing a report to skip over the difficult bits, or the areas you're less sure about (the *Introduction*, for example), and to start with the easy sections (often the *Methods* or *Results* sections). In my view this is a mistake. You should write the report in the same order that you expect the reader to read it in. This is more important than it sounds. You will write a section, having already written the other sections that a reader will have read when they get to that particular one. This means that you're more likely to explain the context correctly and to provide necessary background information in the right place.

Think back to the first chapter of this book. The first two Key Characteristics I talked about were being **purposeful** and **targeted**. I highlighted the following two questions that need thinking about before you start the report writing process:

1. What is the purpose of the report?

2. Who is the target audience for the report?

The answers to these two questions will be provided in the *Introduction* section of the report. So that is the place to start. As the author you must know the answers to these two questions and be able to explain them to the reader. My top tip for producing a good report is to take your time over writing the *Introduction*, ensuring that the key information is included, and don't move on to the next section until you're happy with this one. Get it straight in your mind what information is needed and find out the relevant details. Then sit down and start writing.

What do you write next? Well, I talked in the first chapter about the report being a journey that you need to take a reader on. So the various sections of the report should really be written in order.

Key things to know before you start

The key things you need to know before you start will obviously vary between reports, but the following will be relevant in the majority of cases:

- Client's name and contact details;
- Site name;
- Site Grid Reference;
- Relevant Ordnance Survey maps (or equivalent) and aerial photographs showing the site and the wider area;
- Project description and relevant drawings, including red line boundary plan, and plans showing proposed project;
- Agreed scope of work;
- Specific purpose(s) of report;
- Relevant planning policy (including local planning policy) and legislation;
- Local Biodiversity Action Plan;

- Details of data searches undertaken – organisations contacted and information requested;
- Data search results;
- Details of field surveys undertaken – methods used, dates, times, weather conditions, etc.; and
- Field survey results.

For reports presenting an assessment relating to a proposed development project, such as a Preliminary Ecological Appraisal Report or an Ecological Impact Assessment Report, the following will also be of specific relevance:

- Planning status of project/type of planning application, and planning reference number;
- Any relevant planning history; and
- Outcomes of any consultation, such as pre-application (often abbreviated to pre-app) advice from the Local Planning Authority.

Collating all of this information before starting, or at least those items that are relevant, will ensure that you can write the first few sections of the report with limited distraction. Hopefully this will help with the sections that follow.

In my experience, the key to effective and efficient report writing is building momentum. Get the *Title Page, Introduction* and *Methods* sections written first. There's not much point in writing anything else until you can get these completed, as these will form the basis for what comes next – what information is relevant, how is it phrased, and what are your end goals?

And, once you can look back at some significant progress you will feel as though the report is coming together, and the task that lies ahead should (hopefully) seem less onerous. If you try to produce a report by skipping around between different sections, writing a bit here or there but not finishing any given section, it will be very difficult to track your progress. Of course there will be some minor details that you might need to come back to, which I suggest highlighting in an obvious colour to make sure you don't forget about them, but don't miss out anything fundamental and think you can come back to it – you'll almost certainly end up having to spend more time editing than you otherwise would have.

So, the place to start is, well, at the start. So that's what I'm going to look at in the next chapter.

Chapter 7

First impressions and opening lines

The first few pages of a report should be relatively easy to write for someone that's written several reports before. They can be daunting for someone who hasn't. Either way, we need to take a lot of care over them, because:

1. Some readers might look at the first few pages and form an impression on the quality of work that will follow. Of course everyone knows that you shouldn't do this. Never judge a book by its cover! But on the other hand, everyone does it. Well, everyone will form a first impression at least. If they read further, this might change. However, some won't read further and, in any case, it would be better if readers who decide to delve deeper do so in a positive frame of mind.

2. If the first few bits go wrong, this can have 'knock on' effects later in the report. If the author has not set out the correct information in the *Introduction,* and done so clearly, they are likely to struggle to produce an effective report.

The opening sections of a report can be divided into four parts:

1. *Title Page* or *Cover Page* (this might include an inside *Cover Page* with some details as well)

2. *Contents Page*

3. *Summary*

4. *Introduction*

The *Summary* should be written last. You shouldn't attempt to write the *Summary* until the rest of the report has been completed, and you're happy with the conclusions reached. I'm therefore not going to discuss writing the *Summary* here, as you won't tackle it at the same time as the other parts. Also, there's a considerable amount to say about writing a *Summary*, so I've devoted a whole chapter to them later in the book (Chapter 13).

Let's tackle the other three parts then, in order.

Title Page or Cover Page

There are only a handful of 'rules' for what needs to go onto a *Title Page* or *Cover Page*. This means that there's considerable scope here for a company or organisation (or individual sole trader) to brand their reports as they see fit. Colour, layout and style are all up for grabs. As long as it appears as though it's been professionally produced it doesn't really matter.

There are a few keys pieces of information that must be on the *Title Page*.

Firstly, the report must have a title that goes on the, err, *Title Page*. This is all fairly obvious, but some thought needs to be given to what that title should actually be. Most professional ecology reports will relate to a specific site and to a specific type of report, from the list of those identified in Box 3 (in Chapter 1), or perhaps some other type that I haven't mentioned. The title should include both elements, which should give the reader an immediate idea about the purpose of the report.

In some cases the name of the site might not be immediately obvious. It might be known by more than one name. Its name might change over time (the, not very exciting, Land North of Bridge Street, Bloggshire might become Chaffinch Park, or Babbling Brook, or something else cosy sounding when a housing developer's marketing team start getting involved). It's vital that the name is consistent throughout the report, is correct as far as the client is concerned, given the intended use of the report, and is consistent with the name used by other reports relating to the same project.

So, the title of a report might be, for example:

Land North of Bridge Street, Bloggshire: Great Crested Newt Survey

Chaffinch Park: Ecological Impact Assessment

Bloggshire Brook: Ecological Management Plan

Secondly, there should be some other, very basic, details included on the *Title Page*:

- Author's name and post-nominals
- Name of the author's company or organisation, and contact details
- Client's name (normally a company or organisation, rather than an individual, but can be an individual for some projects)
- Date of issue (at least the month and year, if not the exact date)

Thirdly, there should also be a specific reference number for the report provided on the *Title Page*. This is very helpful for projects where multiple different reports are produced by the same company or organisation, and you can quickly work out which specific report another person is looking at. Importantly, a version number must also be provided. This can be part of the reference number or can be provided as a separate number (or letter) code. It doesn't really matter how you do this, but it's vital that it is done, and that the reference number and version number appear on the *Title Page*.

You may also want to include a copyright notice on the *Title Page* highlighting who owns the copyright, what the report may be used for, and what it may not be used for (as advised in a recent article in CIEEM's *In Practice*).[1] I'm not going to deal with copyright issues any further in this book, partly because they don't really relate to the quality of the report produced, and partly because I'm not a lawyer and I'm therefore not qualified to talk about them. If in doubt I'd advise that you seek specific legal advice.

---

1 Freeths LLP (2020) Copyright Considerations in Ecological Reports. *In Practice* 108, 55–56.

Finally, if the company or organisation producing the report has a Quality Assurance process, then specific details about that will also need to be provided. This might include names and post-nominals of people who have done a proofread or a technical review of the report, for example (see Chapter 16 for more details on proofreading and technical reviews). This information can appear on the *Title Page* itself, or on an inside *Cover Page*.

Contents Page

A *Contents Page* is a list of the headings used in a report along with the page number where each commences. It may also include sub-headings, as well as sub-sub-headings in some cases (although that's often unnecessary). The *Contents Page* allows the reader to navigate easily through the report, finding the specific information of relevance to them. They can also scan through the contents and get a first impression of whether the report includes everything that they would expect to see.

Should all reports have a *Contents Page*? Yes, they should. Well, unless the report is so short as to make it unnecessary.

I was once handed a draft report to review by a relatively junior member of staff when I worked for a large consultancy. The report effectively comprised four pages. The first page was the *Title Page*. The second page was the *Contents Page* (although it only took up a few lines at the top of the page for reasons that are about to become clear; the rest of the page was blank). The third page was the *Summary*, which comprised about one-third of a page of text in a single paragraph. The fourth and final page was the report, including the headings of *Introduction*, *Methods*, *Results* and *Conclusions*, with one paragraph of text under each heading. In this case I didn't think the contents page was needed, as all of the contents either appeared on page 1 (the *Summary*) or page 2 (everything else). Incidentally I'm not convinced the *Summary* was needed in this case either, but that's a topic for another chapter.

This is obviously an extreme case. I don't have a specific rule for how long a report needs to be to justify a *Contents Page*, but I'd have said that a three-page report (including appendices and figures) is too short and a ten-page report is long enough – the cut-off then is somewhere in the middle.

Very few reports will fall into the category of being too short to warrant a *Contents Page*, so the likelihood is that you'll need one. The next question is: what should they look like?

In broad terms it doesn't matter too much, as long as they perform their job. I have only a few general rules for *Contents Pages*.

Firstly, you may have noticed that, in most cases, I've referred to a *Contents Page* in the singular. I regularly read ecological reports with the contents listed over two, three, or sometimes even four pages. My personal view is that this isn't very helpful and is overkill. As I've discussed previously, the human brain will struggle to focus when there's too large a block of text and the same is true when a 'block' of text covers multiple pages.

It isn't the end of the world, of course, if your *Contents Page* appears over lots of pages and, in some cases, it will be necessary (as in this book). However, in a large proportion of the cases where this occurs it is due to there being too many sub-headings (or sub-sub-headings) listed. Often these sub-headings or sub-sub-headings will have the same page reference number – in which case, do you need to list them individually on the *Contents Page*?

Of course your word processing software will probably give you an automatic *Contents Page*, which is very helpful as it will update the page numbering for you. And this will pick out certain levels of sub-headings or sub-sub-headings, which you can dictate. As I've stated previously, I tend to avoid having numbered sub-sub-headings, and I therefore restrict my *Contents Page* to section headings (e.g. *Methods*) and sub-headings (e.g. Desk Study). A sub-sub-heading below 'Desk Study' would not appear on the *Contents Page*. And if I was struggling to get all of the contents on to a single page I would review the various sub-headings for each section and remove them from the *Contents Page* if they all fall on the same one or two pages.

Secondly, as well as the main report headings, sub-headings (where appropriate) and sub-sub-headings (if you must), you should include any appendices and figures (each listed separately). This sounds obvious, but they won't necessarily appear in the 'automatic' *Contents Page*, so you may need to add them manually. Do tables and photos need to be included in the *Contents Page*? I wouldn't have said so, in most cases, although there's bound to be an exception.

Introduction

The first 'proper' text in the report then, other than the *Summary* (which needs to be written last), is the *Introduction.*

There are some key pieces of information that must appear in the *Introduction,* and then a number of other pieces of information that make for a helpful section, giving the reader a broad understanding of what the report is about.

In the first paragraph of the *Introduction* I would always identify the person responsible for writing the report (normally you, as the author), the author's organisation or company, and the person/organisation who has commissioned it (the client). This is essential, and should be in the *Introduction* of every single professional ecological report.

The other key piece of information that you must include in the *Introduction,* is a clear statement of the purpose (or purposes) of the report. Without this the reader may struggle to understand why a particular approach has been adopted, why certain things are considered by the author to be relevant whereas others aren't, and what the end point of the report is – what are the intended outcomes?

The purpose (or purposes) should be provided in a single stand-alone paragraph (i.e. not combined with other pieces of information), to ensure clarity. Where there are multiple purposes (as there will be in many cases) I'd list them as bullet points.

I'm not going to tell you where in the *Introduction* this information needs to sit. I think that will depend on the specific report being written, and how much other background information needs to be included. The *Introduction* needs to flow logically within itself, as well as flowing logically into the rest of the report.

Importantly, the author needs to make sure that the stated purpose (or purposes) also fit the client's brief.

Elsewhere in the *Introduction* I'd suggest having a paragraph that deals with each of the following:

1. A brief description of the site (location and habitats) and its context. You need to remember that most readers won't visit the site, nor know instinctively where it is, and this is fundamentally important to understanding the remainder of the report. It should therefore be set out in the *Introduction*.

2. A brief description of the project that the report relates to. Not all reports will be written in relation to a specific project but many will. This doesn't need to be a detailed description of the project, but enough to explain to the reader what's involved to allow them to fully understand the purpose(s) of the report.

3. Other relevant background information, such as the planning history of the site or reference to other ecological studies. For sites with a considerable amount of history, numerous other previous ecological studies or several related documents, this may take more than one paragraph.

4. The qualifications and expertise of the author (and possibly also anyone involved in undertaking a technical review or final sign-off of the report) to allow an understanding of their competence.

Once you're happy with these opening pages of the report it's safe to move on to the next few sections, normally the *Methods* and *Results*, with a clear purpose in mind. And that's exactly where we're going in the next chapter.

Chapter 8
Getting your facts right

The *Methods* and *Results* sections of a report, along with any related appendices, should be fairly straightforward. These are the sections that are based largely on facts. The *Methods* section will tell the reader what was done, when, where, how and who did it. The results section will describe the outcomes of the study. There is room for some interpretation or opinion in these sections of a report but they should be largely factual.

Methods

The *Methods* section is, normally, one of the easier sections of a report to write because much of the content can be relatively similar in many reports. So, if you're using a report template (which I'll discuss further in a later chapter (Chapter 12)), or borrowing text from a similar previous report as a starting point, with a little editing you can write the *Methods* section pretty quickly. However, there's a danger here, as this can encourage the author to give this section of the report less attention than it deserves. As a result, the *Methods* section can often end up being quite generic, lacking some specific details, and therefore not as **robust** (Key Characteristic 5) or as **clear and precise** (Key Characteristic 10) as it should be.

So, what are the basic rules for writing the *Methods* section? How much detail is needed?

As a whole, a report needs to contain enough detail about the methods used to allow:

1. The reader to draw a reasonable conclusion on the validity of the survey information that the report's conclusions are based upon; and

2. Another ecologist to take that report and repeat all aspects of the surveys in a manner that will get comparable results.

For example, if the weather conditions affect the outcomes of a survey, full details of the conditions during the survey must be provided. If a transect route has been walked during the survey, and the specific route taken could affect the outcomes, then a detailed description of the transect route (or preferably a map or figure showing it) must be provided.

Scope or Rationale

There's a tendency when writing the *Methods* section of any type of ecological report to leap straight into a description of what has been done. This is probably fine for some of the target audience, as they will be ecologists and will implicitly understand the approach to data collection that was needed in a given scenario. However, for some readers it would be helpful for the author to take a step back and explain the reasons why a specific approach has been taken. This explanation can be written as part of the description of desk study and field survey methods, and doesn't necessarily need its own sub-heading. In some cases the survey design will be better explained as part of the description of those methods. However, a specific subsection at the start of the *Methods* section, perhaps entitled *Scope* or *Rationale*, is incredibly useful in setting the scene for the approach generally. This will be particularly useful in reports that deal with several different surveys or numerous different habitats or species, such as Ecological Impact Assessment (EcIA) Reports and Preliminary Ecological Appraisal (PEA) Reports.

This is where the author can introduce the concepts that were part of their consideration when they designed the data collection methods, such as:

- Why was a certain search area used for the desk study?

- Why have some species been focused on but not others?

- Has the approach been discussed with any consultees, such as the Local Planning Authority or a Statutory Nature Conservation Body?

EcIA and PEA Reports will need to make reference to the Zone(s) of Influence in relation to a specific project.[1] This (or these) will dictate, to a certain extent, the study areas for the collection of data. The *Scope* or *Rationale* subsection of the *Methods* is a good place to do that, as it will help the reader understand the description of the methods that follows.

It is important that this part of the *Methods* section is individually tailored for each report. There are some key bits of information that need to be covered in any report of a certain type, but writing this subsection should make the author think about the specifics of the project they are dealing with. This, in turn, will help to stop the *Methods* section from being too generic.

Desk study methods

'Desk study' is the term used to describe the collection of information without visiting a site or its surrounding area (which is described as a 'field visit' or 'field survey'). It is important that a report distinguishes between these two means of data gathering as they provide information that is of value for different reasons. They also have different limitations.

The clearest way to distinguish desk study methods from field survey methods is to give them separate sub-headings. This will be relevant to virtually every single ecological report written. Given the emphasis put on both desk studies and field surveys, and the value of the information gathered from each, a report should clearly state if either of these aspects has not been undertaken, not just that they have. There may be good reasons for not undertaking a desk study in some cases, which I won't go into here as it is a little beyond the scope of a book about report writing.[2] However, the key point is that this should be made explicit in the report. So, even if a desk study hasn't been undertaken, I'd suggest that a sub-heading of *Desk study methods* may still be appropriate, with a **clear** statement explaining that one hasn't been done, and a justification for that. Given the definition of the term 'desk study' that I've set out above, this could include:

---

1 Defined by CIEEM in their Ecological Impact Assessment Guidelines as 'the area(s) over which ecological features may be affected by the biophysical changes caused by the proposed project and associated activities'. CIEEM (2018) *Guidelines for Ecological Impact Assessment in the UK and Ireland: Terrestrial, Freshwater, Coastal and Marine. Chartered Institute of Ecology and Environmental Management*, Winchester. Version 1.1 updated September 2019.

2 See CIEEM (2020) *Guidelines for Accessing, Using and Sharing Biodiversity Data in the UK*. 2nd Edition. Chartered Institute of Ecology and Environmental Management, Winchester.

- A 'data search' with a Local Environmental Records Centre (LERC), or equivalent for information on designated nature conservation sites and existing records of certain species;
- A search for information on designated nature conservation sites using web-based sources;
- A search for existing records of certain species using web-based sources;
- A search for existing records of certain species through a local special interest group;
- A review of aerial photos and appropriate scale Ordnance Survey maps (or their equivalents);
- A review of relevant ecological reports (perhaps produced for the site itself or another nearby location), planning history, or other contextual information which may be available online; and
- Written correspondence received from 'informed individuals' such as the landowner, who might provide details of previous or future management, for example.

A desk study, then, will often go beyond a data search and all aspects included should be reported. And that reporting needs to be **clear and precise**. The information that needs to be provided for each of the examples listed above is set out in Box 5.

Field survey methods

The description of field survey methods will need to answer the following questions:

- What was done?
- Why was it done?
- Where was it done?
- When was it done?
- How was it done?
- Who did it?

Box 5: Desk study information – details to be provided

Element of desk study	Details to be provided
Data search with a LERC, or equivalent	• Name of LERC (or equivalent) • Date of search • Area of search (normally a specific radius around a site boundary or a central grid reference) • Specific information requested
Search for information on designated nature conservation sites using web-based sources	• Name of web-based source • Web address • Date of search • Area of search (normally a specific radius around a site boundary or a central grid reference) • Type of designated site searched for and information recorded for each
Search for existing records of certain species using web-based sources	• Name of web-based source • Web address • Date of search • Area of search (normally a specific radius around a site boundary or a central grid reference) • Name(s) of species
Search for existing records of certain species through a local special interest group	• Name of local special interest group • Date of search • Area of search (normally a specific radius around a site boundary or a central grid reference) • Name(s) of species • Specific information requested
Review of aerial photos and appropriate scale Ordnance Survey maps (or their equivalents)	• Scale of mapping used • Age of maps and aerial photos • Source (such as a specific website)
Review of relevant ecological reports, planning history, or other contextual information which may be available online	• List any sources of information reviewed with a full reference
Written correspondence received from informed individuals	• Reference as appropriate

For each field survey undertaken, such as an initial habitat survey, a bat activity survey, or an assessment of the suitability of trees or buildings for roosting bats, for example, some key pieces of information will need to be provided.

Firstly, explain **what** specific type of survey was undertaken. Some ecological surveys will have names used in survey guidance documents, such as 'Phase 1 habitat survey', 'NVC survey', or 'Preliminary Roost Assessment'. Others will have widely accepted names that help to distinguish one type of survey from other types, such as 'Dormouse nut search' or 'Dormouse nest tube survey'. These 'standard' names should be used wherever they exist. Care needs to be taken when selecting and using these terms to ensure that the reader isn't misled into assuming a specific approach has been taken when it hasn't. For example, a survey of a site to identify the birds that are breeding on it could obviously be called a 'breeding bird survey'. However, the British Trust for Ornithology (BTO), Joint Nature Conservation Committee (JNCC) and RSPB use a survey known as a Breeding Bird Survey (BBS) to monitor bird populations; this is based on a specific number of survey visits and design of survey transects.[3] This could obviously cause confusion.

Secondly, the report author needs to make it clear **why** a survey has been done, and why a particular survey approach has been selected. This issue has already been discussed under the sub-heading of *Scope* or *Rationale*. There's no need to repeat the explanation of the rationale for a survey in the description of the specific methods if it has already been given earlier in the report. However, the specific reasons why a given approach has been adopted will often sit better within the description of methods, rather than in a more general subsection on the overall scope.

One of the key reasons why a specific approach has been used is often that it is recommended in the relevant good practice guidelines. Reference to those guidance documents will therefore be an essential part of the description of survey methods. In many cases it won't be sufficient to simply state that an approach taken accords with a specific set of guidelines – the report will need to provide enough detail to allow someone reviewing the report to be satisfied that it does. This will be particularly important where there is more than one specific approach, with different options considered appropriate in different cases. These different options might relate to different levels of

3 See https://www.bto.org/our-science/projects/bbs/research-conservation/methodology.

survey effort, or completely different techniques. The selection of one option over another will need to be explained.

The specific location or locations **where** each survey took place must be clearly stated. In most cases this will require the inclusion of a suitable map or figure, showing transect routes for example, or vantage points, pond locations, or parts of a site where surveys for a particular species were carried out. Further discussion of the use of figures is provided in Chapter 11.

In all cases a report should state the exact date **when** a survey was undertaken. There is no reason not to provide this piece of information, and it will be relevant in many cases to the repeatability of the survey.

Certain surveys need to take place at specific times of day, such as bat activity or emergence/re-entry surveys, reptile surveys, bird surveys and amphibian torchlight surveys. Other surveys will require the surveys to be undertaken over a certain period of time to ensure a sufficient level of survey effort. In both cases the start and finish times of the survey need to be provided.

For some other surveys there's no need to state the specific times when the survey took place, as it shouldn't make a material difference (assuming that they were conducted during the hours of daylight unless dictated otherwise by the recommended timings of the survey technique). There will be situations where surveys that are normally conducted during daylight hours have to be undertaken at night, such as where they are required on live railway lines and are restricted to night-time when the line can be temporarily closed. In these cases the specific timing of the survey must be stated, and the potential limitations discussed (which we'll come on to later in this chapter).

Fairly obviously, the description of survey methods will need to give details of **how** each survey was undertaken. Again, this will need to refer to the relevant guidance documents, where available, and provide enough detail to allow another ecologist to repeat those surveys.

And finally, it's important to state **who** undertook each survey. The competence level of the surveyor is fundamentally important in making an assessment of the validity of the survey data, so their name and relevant qualifications (including details of survey licences held, where relevant) should be included.

Limitations

All methods of data collection have their shortcomings. The validity of species records provided as part of a data search can be difficult to gauge – it depends on the competence level of the person who provided the record and whether it has been corroborated. And a lack of desk study records of a given species doesn't mean that a species is not present (absence of evidence is not the same as evidence of absence). A field survey will often take place over a prescribed number of visits and is therefore a sample of what was present at specific locations during those visits. Just because a particular species wasn't recorded during those survey visits doesn't necessarily mean it wasn't there on other occasions, or couldn't appear in the future. Field survey results will therefore make reference to 'likely absence', as absence can rarely be guaranteed.

The general shortcomings of the approach to data collection need to be made explicit to the reader. However, in my view they are different from specific survey limitations.

I've read ecological reports that provide a description of these general shortcomings under the sub-heading of *Limitations*. I don't think there's a problem with doing this, provided it is done in the right way, giving a general health warning rather than acting as a disclaimer, and that such issues are given fairly brief treatment in the report. The alternative is to make the general shortcomings of data collection **clear** through the use of careful language around the way that survey results are described and interpreted in the report.

What will definitely need to appear in the *Methods* section of a report is a description of any specific factors that may have affected the outcomes of the survey(s) in that particular case. These are the sorts of factor that I would consider, and describe as limitations. They could include, for example:

- Poor weather conditions during the survey that may have affected the survey results;
- Lack of access to a part of the site or area around the site that would otherwise have been included in the study area;

- Surveys undertaken outside of the optimal survey season or outside of the times of day that would normally be expected for that survey type;
- Fewer visits or lesser survey effort undertaken than is the recommended minimum level of effort to achieve the desired outcomes;
- Following a survey approach that does not fit with the one recommended in relevant good practice guidelines; or
- On-site management practices making it impractical or inappropriate to undertake the widely accepted approach.

Personally, I would distinguish these from the general shortcomings of all means of data collection by thinking about another ecologist trying to repeat the surveys in order to get comparable results. For example, if the published good practice guidelines on surveys for a given species recommend four survey visits over a specific period, then the fact that this is a sample is relevant to the interpretation of the information, but not to the individual attempting to repeat that survey. If only three visits could be undertaken instead of four, or there were access restrictions, or the weather wasn't ideal during one specific visit, then these will be relevant factors to report.

These limitations on the surveys can be described under a sub-heading of *Limitations* or alongside the description of each survey method, perhaps under a sub-heading for each survey type. Importantly though, it isn't enough for a report to simply identify these limitations, it must go on to explain their significance (or not) in the context of the purpose of the report. The same limitation, such as restricted access, might be completely inconsequential to the outcomes of the assessment presented in the report in one context, but fundamentally important in another. The limitation needs to be identified in both cases, and the consequences explained in the *Methods* section, with this then picked up and discussed as appropriate in subsequent sections of the report.

Particular care needs to be taken when writing this section of a report as a failure to explicitly identify a limitation could have serious consequences for any subsequent reliance on the conclusions of the report, and this may be seen as an attempt to deliberately mislead. If in doubt about whether to include a possible limitation it will always be better to include it and discuss it openly than to leave it out of the report. This is a key part of ensuring 'full disclosure', as highlighted in Section 6.7 of the British Standard on Biodiversity.[4]

Some ecologists use the term 'survey constraints' to refer to the survey limitations. The problem with this is that the term 'constraint' is also used to refer to an ecological feature that might constrain how a proposed development is designed or constructed. I'd therefore suggest avoiding the use of the term 'survey constraints' and stick with 'survey limitations' instead.

Assessment methods

For certain ecological reports (EcIA Reports, for instance) there will be an element of assessment. The approach to the assessment also needs to be explained in the *Methods* section. I won't go into detail about that here, as it's moving into the area of technical details associated with undertaking the Ecological Impact Assessment process. The key point I want to make, though, is that it needs to be covered. And having read many EcIA Reports over the years I know that it is often forgotten about.

Using appendices for details of survey methods

This is discussed in detail in Chapter 11. In general I'd encourage authors to move as much of the description of **when** and **how** a field survey was done, along with **who** did it, into appendices. This helps to balance the report more towards your average reader. Those who want to look at the specific dates, times, weather conditions and personnel for each survey visit can be directed towards that information, without the flow of the report being interrupted by large tables full of these details.

Of course, it depends on how much information falls into this category. If there isn't a great deal of it, then it can be better to leave it in the main *Methods* section of the report, otherwise you can be unnecessarily complicating life for some of your readers by making them cross-refer to a separate part of the report, without a great benefit to the rest of the readership.

---

4 The British Standards Institution (2013) *Biodiversity – Code of practice for planning and development. BS42020:2013*. BSI Standards Limited.

The main *Methods* section should always include the elements of **what** was done and **where** it took place, as knowing this will be material to understanding the report. And I'd suggest that **why** a certain approach was taken should also be included in the main report and not moved into an appendix, including references to good practice guidelines, as necessary, to explain the rationale.

Limitations should always be dealt with in the main body of the report in my view, as to move them to an appendix could be seen as trying to hide them.

Results

As with the *Methods* section, the description of the outcomes of the desk study and field surveys undertaken needs to be **robust** and **precise**. All of the information collected as part of the various studies should be included in full, with a couple of notable exceptions:

1. Data search results. A request to a Local Environmental Records Centre for existing records of protected and priority species, for example, can often generate multiple pages of data. Personally I don't think that this all needs to be included in a report. Indeed you could argue that it shouldn't be reproduced in a report as there may be copyright issues associated with doing so. The data should be reviewed and any relevant records extracted, and they should appear in the report. The whole dataset does need to be made available on request to allow someone reviewing the report to check for themselves that the relevant data have been extracted and presented correctly.

2. Automated bat detector results. The results from an automated bat detector could well comprise thousands of separate passes by bats; many of these will be from different species and at different times of the night. I've never been convinced that there is a need to present all of this information in full in a report. Instead I think it is appropriate for the report author to summarise the results in a manner that informs the assessment. In most cases, the number of passes per night by different species, in different parts of the site, will be of relevance, as it will give an indication of relative levels of bat activity. The results can therefore be summarised into a simple number of passes by each species on each night of survey. In some cases the time of the passes will be relevant, as

this might dictate proximity to roost sites, or the relative importance of the site. In that case the times of each pass, or the times of the earliest and latest pass should also be provided. This will be more relevant for some species than for others. Again, the original dataset should be available on request.

There will, I'm sure, be other examples of scenarios where the results can be summarised rather than presented in full in a report. However, these are going to be the exception rather than the rule.

It's worth highlighting here that EcIA Reports tend not to have a section titled *Results* at all. The outcomes of the surveys are normally used to inform a description of what is termed the 'Baseline conditions' for the assessment, and this is only normally described for sites, habitats and species of relevance to the proposed development. This means that survey results that don't relate to relevant sites, habitats or species won't necessarily be described or made reference to in the main body of the report. They should still be included in the report elsewhere for completeness, such as in an appendix.

Desk study results

There will be many different outcomes of a desk study, and how and where you report them depends, to an extent, on the type of report you are writing and the specific relevance of the information.

An EcIA Report or PEA Report will need to include information on the designated nature conservation sites within the Zone of Influence of the proposed development project. The information on designated sites will need to include the following as a minimum for each site:

- Status of site/type of designation;
- Reasons for designation;
- Distance from proposed development site; and
- Direction from proposed development site (north, north-east, east, etc.).

In some cases, more detailed information about the interest features that the site is designated for will also need to be provided.

The information for the relevant designated sites could usefully be presented in a table.

It may be that there are numerous designated sites within the search area chosen for the data search, but the majority of these are clearly not directly relevant. In this case the table of information relating to all designated sites could be included as an appendix, with the ones of relevance discussed further in the main body of the report.

Most ecological reports will include information on a given species or habitat that comes from desk study as well as information on the same species or habitat that comes from field survey. It is important to distinguish between these two basic sources of information when writing a report and there are two ways to achieve this.

One option is to have separate subsections for desk study results and field survey results. This works well in a report dealing with a single species or species group, such as a reptile survey report for example. It doesn't work so well in reports that deal with multiple habitats and species, such as an EcIA Report or PEA Report. The reason for this is that desk study results and field survey results often need to be related to each other.

The alternative approach is to discuss them in the same part of a report, under the sub-heading of a given habitat, species or species group. This often helps the reader understand the outcomes and can also help the author in writing the relevant sections. Information gathered through a desk study will often assist with explaining the context of a field survey result, and the findings of a field survey will often enable the effective resolution of an issue raised by a desk study record. In this case the author needs to take care to ensure the reader understands whether the information comes from desk study or from field survey (and indeed the data source, or specific field survey, that generated the information). For example:

> *The data search confirmed that there are existing records of Japanese knotweed, an invasive non-native plant species listed on Schedule 9 of the Wildlife and Countryside Act 1981 (as amended), at a location approximately 1km north of the site boundaries (Ordnance Survey Grid Reference SX123456). There are no existing records from the site itself. This plant was not recorded during the extended Phase 1 habitat survey of the site, and is considered likely to have been recorded if it were present. It is therefore considered likely to be absent from the site.*

Here's another example:

> *The site was assessed as providing optimal habitat for reptiles, as it comprised an extensive area of unmanaged grassland with scattered scrub on a south facing slope. An adder was incidentally recorded basking within the site during the extended Phase 1 habitat survey; a targeted survey for reptiles has not been undertaken.*
>
> *The data search confirmed the presence of adders, grass snakes, common lizards, and slow-worms at Snake City Site of Special Scientific Interest (SSSI), approximately 2km to the west of the site, and at Lizard Lounge County Wildlife Site (CWS), approximately 3km to the east of the site. There are no existing records of reptiles within 5km of the site outside of the designated sites referred to above, and the landscape to the north and south of the site generally comprises intensively farmed arable land likely to support relatively few patches of good quality habitat for reptiles.*
>
> *The site may therefore provide an important link between reptile populations associated with Snake City SSSI and Lizard Lounge CWS …*

It is difficult to generalise about the desk study data that need to be provided with the exception of designated sites, but it will normally fall into one of several categories:

- An existing record of presence. This tells you exactly that. In this case the specific location relative to the site, or within the site will be relevant and should be shown on a map or accurately described.[5] The date of the record will also be relevant.[6]
- An existing record of presence within the area surrounding a site that triggers the author to consider the likelihood of presence within the

5 In some cases it is considered appropriate to keep specific locations confidential, such as the location of a badger sett or the nest of certain bird species – in which case either the location will need to be described in vague terms or, if that isn't acceptable in the context of understanding the report, the report itself or certain parts of it will need to be clearly labelled as confidential and kept out of the public domain.

6 There will be cases where existing records are so old that they are no longer relevant to the current situation, but there is no specific 'cut-off' date, and if the report author chooses to disregard a record on the basis of its age (or any other factor) they will need to clearly explain why they've done so.

site and or Zone of Influence of a project, or provides useful context about the abundance or distribution of the habitat or species concerned, or the relative importance of the site. Again, the specific location and date of the record will be relevant, as will other factors such as habitat preferences.

- Contextual information that aids the understanding of trends or patterns, or likely future outcomes. Precise locations are likely to be of lesser relevance in this case.

Field survey results

The means of presenting the results of field surveys will generally fall into three categories:

1. Tables of data
2. Descriptive text
3. Maps, figures, drawings or photos

Tables are useful to present significant amounts of data, and are the subject of a later chapter in this book (Chapter 11). In the context of field survey results they might be used, for example, to:

- Show the numbers of great crested newts, subdivided into males, females and subadults or larval stages, recorded on each survey visit to a pond, using each of a number of different techniques;
- List the species recorded as part of a bird survey of a site during the breeding season, identifying the species observed on each visit, whether activity indicative of breeding was recorded, and an estimate of the number of pairs; or
- Summarise the results from an automated bat detector at a given location, showing the number of passes by each species on each night of survey.

Short tables (anything up to half a page) can be included in the main body of the report, but longer tables should be included in the appendices. In most cases the results presented in tables, or at least a summary of those results, will also need to be described in text.

Some results are really only presented as descriptive text, such as:

- Descriptions of the habitats present at a site; or
- Descriptions of the suitability of those habitats to support a particular species.

In some cases the results of a survey might be better summarised in words rather than in a table of data, such as where there were multiple survey visits on multiple occasions, but no (or very few) records of the target species. In this example a large table with lots of zeros in the rows and columns is unlikely to be a helpful way of presenting the information and is probably unnecessary.

When writing the *Survey results* section of a report, it is important to stick to the facts as far as possible. Some of the results will be an opinion of sorts, such as where a habitat has been classified into a particular category. Others may disagree with the classification. It is therefore important to provide the evidence that backs up the classification that has been chosen.

Many surveys will generate information where the locations of those data points are relevant, such as the location of a pond, a tree containing features suitable for use by roosting bats, or a badger sett. This information should be presented in a map or figure (see Chapter 11 for a discussion of the use of figures). This won't be the case for all of the information recorded during a survey – in some cases the specific locations are irrelevant. For example, the exact location where a brown hare was seen will not normally need to be shown in a report – it's presence in a particular habitat type within the site is all that really needs to be recorded. Having a figure showing lots of data points where the specific location is of no real relevance can be unhelpful. Some thought will therefore need to be given to the most appropriate way of presenting the information.

There are other visual forms of presenting survey results, such as photographs, which I'll also discuss further in Chapter 11.

Dealing with incidental or 'non-target' information

A survey will often generate a result that is not directly relevant to the purpose of the survey. The information might, however, still be of relevance in another context and should therefore be recorded and presented in the report. It should, however, be given relatively brief treatment in comparison with the target of the survey.

For example, a range of non-target species are often recorded under refuges used for reptile surveys, such as field voles, bank voles, common toads and common frogs. Although not the specific target of the survey, the presence of these species may nevertheless be of relevance. They can be included in any tables of results as 'incidental records', or 'other species recorded'.

Sometimes a species or feature is recorded that, although not the specific target of the survey, is clearly something that the surveyor feels that they need to report as they know that it is of relevance to other studies being undertaken. An example might be the discovery of a badger sett in the banks of a watercourse during a water vole survey. The surveyor knows that badger setts are protected and should therefore consider it appropriate to report its presence – either in their report or through another appropriate means.

Chapter 9

So what does all this mean?

Once you've written the *Methods* and *Results* (or equivalent) sections of a report, you'll then need to move on to interpret the information you've collected.

Virtually every single professional ecological report will need some form of interpretation, setting out the implications of the findings, in most cases, for a specific project. There will be exceptions – Habitat Management Plans, for example, might not require much (if any) interpretive text.

Dependent on the type of report and its purpose, the 'interpretation' might fall under a number of different section headings, such as:

- *Discussion* (normally used for survey reports or research reports);
- *Recommendations* (normally used for survey reports and PEA Reports);
- *Impact assessment*, *Mitigation* or *Assessment of effects* (normally used for EcIA Reports);
- *Conclusions* (most professional ecological reports should include a *Conclusions* section).

These sections of a report are often the most difficult to write. They require the author to use their experience and understanding to support an opinion and therefore to demonstrate sound professional judgement (see Box 2 in Chapter 1). In some situations there will be lots of advice or guidance available to help with this; in other cases there will be much less. In Chapter 1,

I set out a list of 10 Key Characteristics for reports. I've referred back to these in several of the intervening chapters, as some are specifically relevant to different topics, or to certain sections of a report. For the 'interpretation' sections of a report, such as you might write under the example section headings above, pretty much all of these Key Characteristics will be relevant. For this reason I've listed them again below:

1. **Purposeful** – has clear aims and objectives that meet the expectations of the intended target audience, and that it delivers against;

2. **Targeted** – is written with its target audience in mind;

3. **Well structured** – has a logical flow that makes it easy to follow;

4. **Transparent and truthful** – is open and honest about the data presented, the sources of data, any limitations to collecting the data, any interpretation of the data, and whether a statement is a fact or an opinion (see Box 2 for a definition of these terms);

5. **Robust** – is based on sound data which is sufficient and appropriate to support the purpose;

6. **Justified** – provides suitable evidence for any conclusions reached and recommendations made;

7. **Written by a competent person** – the author is suitably qualified to make the judgements contained within the report;

8. **Impartial** – is not biased towards a point of view that either benefits or disadvantages any stakeholder, including the client;

9. **Proportionate and concise** – provides an appropriately balanced treatment of the issues, with more 'air time' given to those that are more important or more complex, without providing excessive amounts of unnecessary information;

10. **Clear and precise** – sets out the information in a manner that is easy to understand, unambiguous and with attention to detail.

I've already discussed being **purposeful** and **targeted** (in Chapter 1), so I won't go into any further detail here, except to say that it's worth an author reminding themselves of the specific purpose (or purposes) of the report, and who the target audience is, whilst writing the 'interpretation' sections. It's easy to write a purpose in the *Introduction*, then get stuck into *Methods* and *Results*, and have completely forgotten about it by the time you come to write these later parts of the report.

I've also discussed **competence** in an earlier chapter (Chapter 2) so, again, no need to revisit that here in any detail. However, quite clearly the more reliant a report is on an author's personal experience as evidence (for example, where there is little or no existing published advice or guidance) the more specifically relevant their competence will be in terms of the 'interpretation' sections. And, of course, if the document being written is a piece of good practice guidance, the issue of competence will be of even greater relevance.

Finally, I've also already addressed the requirements for reports to be **robust**, by being underpinned by appropriate data, in the previous chapter.

So, let's look at those remaining six Key Characteristics in a little more detail, thinking specifically about the 'interpretation' sections of a report.

Well structured

The first decision to make is what section headings and sub-headings to use. As with all reports in general, the way the report is divided up is important, to ensure that it flows logically and avoids unnecessary repetition. To an extent this can be standardised, so that reports of a certain type always follow the same pattern, with a consistent set of headings and sub-headings. However, it's easier to have a standardised structure for the earlier sections of the report than it is for these later 'interpretation' sections.

One of the options here is to follow the sub-headings that have been used in the previous (normally the *Results*) section. This is logical and will help the reader find the relevant bits of information. In some cases, though, it might be beneficial to switch to a different set of sub-headings – one more tailored to how the reader will use the information. For example, a PEA Report will normally include a section that provides recommendations for a client in relation to a certain project. These will include:

- Recommendations for how they might amend the design of the project;
- Recommendations for further ecological surveys necessary to underpin a planning application;
- General recommendations for mitigation measures likely to be required (subject to a detailed design); and
- Recommendations for enhancement measures that could be considered.

Separating the various recommendations into subsections such as those just described, effectively grouping issues together that have similar implications for the client, will be more useful than subsections based on the habitat or species that each relates to (as might be the case for earlier sections of the report). In this latter case issues with major implications, which the client needs to clearly understand, can become lost amongst more trivial ones.

The structure of the 'interpretation' sections therefore warrants a little consideration. A 'one size fits all' approach is unlikely to work.

Secondly, it's important that the text is structured under each heading or sub-heading. In the first chapter I talked about taking the reader on a logical journey from the initial purpose or purposes of the report to its conclusions. The same is true, on a smaller scale, for each section or subsection in the report. This is particularly important for the 'interpretation' sections of a report where the author will need to lead the reader from a single finding to an inference; this is likely to need support from reference sources. The reader will have to be introduced to facts, inferences and any supporting references in a logical way, allowing them to follow a series of simple steps and arrive at the inference that is being made. The reviewer of a report will want to see those steps and may challenge some of them. Part of a report being **well structured** then is about ordering the information that is being presented so that the reader can follow the logical steps. Ensuring that the report can withstand the reader challenging those steps is part of what makes it **robust**; we're going to come onto this next.

Transparent, truthful and justified

In Chapter 4 I discussed the difference between fact and opinion, and the importance about being **clear** in a report as to whether a statement is one or the other. Being **transparent and truthful** then is of particular relevance to the 'interpretation' sections of a report. You are expected to set out the implications of your findings, and this necessarily requires an opinion to be expressed. It should be based on a combination of facts, published evidence (such as guidance documents, or peer-reviewed research articles) and, in some cases, unpublished evidence or experience (but see Box 2 in Chapter 1, and Chapter 4 in relation to this). This is effectively part of demonstrating sound professional judgement. And the report therefore needs to be open and honest about the sources of information being used, particularly where they are being used to support an opinion.

The importance of supporting statements of opinion with appropriate reference sources is a part of making sure that a report is **justified**. This issue was also discussed, in principle at least, in Chapter 4. The difficulty, though, is in how you apply this in the context of different reports. Here are some examples.

1. *Assessing the likely presence (or absence) of a species, based on equivocal survey results*

   It's a fact of an ecologist's life that field surveys don't always provide definitive results. Even when a survey is conducted at the right time of year and during appropriate weather conditions, and for a species that has characteristic field signs, it's not always guaranteed that the survey will be definitive. Nevertheless, a professional ecologist might think it highly likely that a particular species is present or absent, based on various characteristics of the site, such as its location and the habitat it supports, or characteristics of the species concerned, such as how likely a survey is to find evidence. The ecologist will need to express an opinion, but they'll need to be careful to justify that opinion. For example:

   > *Field signs indicating the <u>possible</u> presence of water voles were recorded during the survey. However, the field signs recorded (feeding remains) were not sufficiently distinctive to allow the presence of water voles to be conclusively confirmed. Other species of vole leave very similar*

field signs (see Ryland and Kemp 2009).[1] *Given the time of year when the survey was undertaken other, more definitive, field signs would normally be recorded during a survey; no such signs were recorded in this case. However, the site provides suitable habitat for water voles, and the desk study confirmed that this species is present on the same watercourse within 500m of the site (see Section 3 of this report). In the author's experience it is therefore considered possible that water voles are present at relatively low densities. This report has therefore been based on a precautionary assumption of the presence of water voles.*

The desk study confirmed the presence of slow-worms within 1km of the site, associated with a nature reserve (off site) comprising unmanaged, tussocky grassland (see Section 3 of this report). The site provides suitable but relatively poor habitat for slow-worms, comprising intensively managed arable fields with limited margins, which is not a habitat type generally considered to be 'favoured' by reptiles (as opposed to 'unintensively managed farmland' which is favoured – see Froglife 1999).[2] *In addition, the wider landscape surrounding the site is intensively managed farmland, as is the land between the site and the nature reserve referred to above. The site is therefore not well linked to the nature reserve where slow-worms are known to be present, and nor is it well linked to other patches of habitat likely to be of value for this species. The targeted reptile survey of the site did not record any slow-worms or any other species of reptile. It is therefore considered likely that slow-worms are absent.*

2. *Assessing the suitability of habitat to support a particular species*

Determining habitat suitability is highly subjective and will require a significant amount of experience on the part of the person making the assessment. Yet all too often reports are written in a manner that makes it sound as though this is a fairly cut and dried assessment, with little room for disagreement. Statements like this, for example:

---

1 Ryland, K. and Kemp, B. 2013 Using field signs to identify water voles – are we getting it wrong? *In Practice* 63: 23–25.

2 Froglife (1999) *Reptile survey: an introduction to planning, conducting and interpreting surveys for snake and lizard conservation. Froglife Advice Sheet 10.* Froglife, Halesworth.

The site provides suitable habitat for hazel dormice.

The site is unsuitable for hazel dormice.

The site provides suboptimal habitat for hazel dormice.

These statements may all be true in different cases, but they need to be supported with evidence.[3] For example:

Hazel dormice are known to utilise a variety of habitat types, including woodland, scrub and hedgerows, and exhibit a preference for a dense and well connected understorey, and high plant species diversity, as this provides a variety of food sources across the year (Bright et al. 2006).[4] *The site supports semi-natural woodland, with a dense understorey and tall (8–10m), wide (3–4m), species-rich hedgerows, which are well linked to each other, as well as to other similar habitats in the wider landscape surrounding the site. The site is therefore considered to provide suitable habitat for hazel dormice.*

The site comprises grassland fields, which are intensively grazed, subdivided by fences with occasional patches of nettles. None of these habitats provide the structural diversity and variety of food sources required to support hazel dormice (see Bright et al. 2006). The site is therefore considered to be unsuitable for hazel dormice.

The site itself does not support any hedgerows or areas of woodland, although such habitats are present throughout the wider landscape surrounding the site. The site does support small patches of bramble scrub, which may be used by hazel dormice associated with off-site areas of woodland and hedgerows, providing a suitable structure for nesting (particularly from late summer onwards when the vegetation is at its densest) and a valuable foraging resource in autumn, when the bramble is fruiting (Bright et al. 2006). Overall, therefore, the site does

---

3 In these examples I have referred to the 2006 Dormouse Conservation Handbook, published by English Nature/Natural England, as the most relevant reference source. However, at the time of writing, a revised Handbook, due to be published by the Mammal Society, had been drafted and is likely to replace the 2006 version in 2020 or 2021 as the most relevant (and up-to-date) reference source.

4 Bright, P., Morris, P. and Mitchell-Jones, T. (2006) *The dormouse conservation handbook. Second edition*. English Nature, Peterborough.

not support sufficiently diverse and well structured habitats that would be considered to be 'optimal' for hazel dormice, but it is nevertheless considered suitable.

3. *Assessing geographic scales of importance, such as in an EcIA Report*
The Ecological Impact Assessment (EcIA) process requires an assessment of the scale of importance of ecological features (designated sites, habitats or species populations). There are some different approaches to how you might tackle this, which I won't go into here, but this effectively relies on the availability of contextual information. Contextual information allows the author to compare, for example, a species population within a site with populations of that same species within defined geographical areas, such as the country or county that the site is located in. The main difficulty with this, in most cases, is the general lack of contextual information. In other cases there may be multiple sources providing different information. This makes it important to provide a reference source for the contextual information being used, and to make it **clear** where assumptions are being made.

It's also important that the report is written in a way that allows the reader to follow the author's reasoning, by breaking the assessment down into a series of logical and justifiable steps.

I've often read statements in EcIA Reports in relation to this that are completely unsupported by any evidence or explanation of how the assessment has been made. For example:

The population of common toads present within the site boundaries is of 'Local' importance.

If this were rewritten with appropriate supporting evidence, it might read something a little more like this:

Common toads are listed as a species of principal importance for the conservation of biodiversity in England under Section 41 of the Natural Environment and Rural Communities Act 2006. The population of common toads present within the site boundaries is associated with the breeding pond located on the site's northern boundary, at Ordnance Survey Grid Reference ST123456. Surveys of the pond during March 2020 recorded a maximum count of 26 adult toads using the pond. This is significantly lower than the number required for designation as a 'Key Wildlife Site' – criteria S3.2 of the Gloucestershire Key Wildlife Sites Handbook (Part 2) for designation of such sites includes 'any pond where 100 or more adult common toads are observed in early spring'. On this basis the site's toad population is considered to be smaller than would be required to assign it 'County' scale importance. Nevertheless, it is still considered to be of importance, given the status of toads in the local area, and has therefore been assigned 'Local' scale importance in this case.

4. *Assessing likely impacts and the likely effectiveness of a mitigation measure*
   We're now into a hugely subjective area but one that is dealt with in large numbers of professional ecological reports. There are, of course, reference sources that will help an author to identify the sorts of impact that might affect a particular ecological feature, whether a habitat, a species population or a designated site, and then to help them assess those impacts. The report needs to be **clear** about the reference sources being used.

   Some ecological features will not have a published reference source that allows the author to justify focusing on specific impacts and discounting others; the author will need to determine the appropriate impacts themselves. Where this is the case it needs to be made **clear**.

   The process of Ecological Impact Assessment will require the author to go further than simply identifying impacts. They will need to consider the sensitivity of the feature to a given impact and determine whether that impact will change the feature in a significant way – normally by determining whether there would be a change in conservation status or not. This will require the use of appropriate reference sources, otherwise the author is simply stating a subjective opinion of what change might occur and a further subjective opinion of whether that change should be considered to be significant or not. I don't want to get too drawn into the process of Ecological Impact Assessment here – that's a book in itself – but there's an important point to be made about how we present the outcomes of that process in an EcIA Report.

   EcIA Reports will also need to address the issue of how likely a mitigation measure is to be effective. As with identifying and assessing likely impacts, this is an area that invites opinion, and requires evidence from appropriate reference sources to ensure that the opinion is supported. What might seem obvious to an ecologist will not necessarily be to someone who isn't an ecologist. And for EcIA Reports aimed at a wide target audience, many of which won't be ecologists, this will need to be considered.

For example:

a. I know that otters are sometimes killed when trying to cross roads; a new road therefore represents a hazard to them. I probably don't need to justify this assertion as it is fairly obvious and uncontroversial.

b. I also know that otter deaths often occur at locations where roads cross watercourses. That seems pretty obvious as well and is supported by published research (Philcox *et al.* 1999).[5] However, there have been numerous examples of otters being killed at locations well away from watercourses. Therefore, to ensure that the assessment is 'balanced', it will be important to highlight this with an appropriate reference source (such as Chanin 2006).[6] I can't confidently assume that otters will only encounter the new road at locations where it crosses a watercourse.

c. There is evidence that certain designs of crossing structures make it more difficult for otters to pass safely under a road than others. There has been some published research into this issue (Grogan *et al.* 2001).[7] There may also be published advice on the design of crossing structures.[8] I will need to consider the design of the structures proposed against the findings of the research and any current published advice to allow me to assess the likelihood of otters passing under the road safely, versus the likelihood of them trying to cross over the road and therefore subjecting them to the

---

5 Philcox, C.K., Grogan, A.L. and Macdonald, D.W. (1999) Patterns of otter *Lutra lutra* road mortality in Britain. *Journal of Applied Ecology* 36: 748–762.

6 Chanin, P. (2006) Otter road casualties. *Hystrix, the Italian Journal of Mammalogy* 17(1): 79–90.

7 Grogan, A., Philcox, C. and Macdonald, D. (2001) *Nature conservation and roads: advice in relation to otters*. Wildlife Conservation Research Unit, Oxford.

8 In 2020, Highways England revised its *Design Manual for Roads and Bridges* which, until then, had contained detailed advice on the design of crossing structures in relation to otters.

risk of being killed. These documents will need to be referenced appropriately.

d. When otters move across the landscape by following watercourses they are not always swimming within the channel – they run along the banks and are particularly likely to do this when travelling upstream against the current. It logically follows that otters will be more likely to move on land during times of high flow and there is evidence to support the assertion that otters are more likely to be killed during such times (Philcox *et al.* 1999), although experience has shown that not all otter road deaths can be explained on that basis alone. It also follows that, when rivers are in spate, the precise location at which an otter will encounter a road is likely to be less predictable than under 'normal' conditions.

e. The references cited above are relatively old. More recent evidence may be available, which should be reviewed. I'm aware that many previous studies have struggled to draw firm conclusions about the effectiveness of different designs of crossings in terms of their use by otters, due to the difficulty of determining what an otter does when it encounters a structure and that this might occur relatively infrequently.

What I've done in paragraphs (a) to (e) above is to highlight some of the difficulties that a professional ecologist might face when tackling the issue of assessing likely impacts and determining the effectiveness of mitigation. The main problem being that there is often a lack of conclusive evidence. This must be made **clear** to the reader. An EcIA Report must not **deliberately** mislead a reader into thinking that there is certainty about something when there isn't. That should go without saying. However, the author also needs to do everything possible not to mislead a reader **accidentally** as well. The more **transparent** a report can be made in this regard the better in my opinion.[9]

---

9 It is worth highlighting here that, at the time of writing, the Mammal Society has started drafting an 'Otter Mitigation Handbook'. This should provide an up-to-date and relevant reference source for discussing this particular issue, once published. Importantly, this document will also need to be clear about the evidence used to underpin any guidance or advice it provides.

What might I have written in the impact assessment for the new road, in terms of the likelihood of otters being killed crossing that road? I'm not going to give you a full set of text here, as it's impossible to generalise about a hypothetical situation, and it would run the risk of someone copying it, possibly even using it out of context. I'll therefore answer the question by simply saying that I'd have touched on all of the issues that I identified in paragraphs (a) to (e) above.

Impartial

The example I've just given, relating to otters and roads, highlights the importance of ecological reports being balanced in their treatment of potentially controversial issues. This is vitally important to the ecology 'industry'. Professional ecologists must remain **impartial.**

The way that we recognise and deal with ecological issues through the planning system is, to a large extent, reliant on this impartiality. Those ecologists contracted by developers to write reports on their behalf (an ecological consultant) must be careful not to mask or downplay important issues. Those working for Local Planning Authorities or representing other interested parties also have a part to play in this. They must be careful not to overstate the importance of issues of limited relevance. In some cases one party might keep the other in check. But that doesn't work very well in parts of the country where the Local Planning Authority doesn't employee an ecologist.

It's also worth noting that CIEEM, the professional body that many of those working as ecologists belong to, specifically mentions this in its Code of Professional Conduct. Those professional ecologists who are members of CIEEM have signed up to this 'Code' which includes:

> *As a member of CIEEM I shall … exercise sound professional judgement in my work, identifying clearly the limitations and applying objectivity, relevance, accuracy, proportionality and impartiality to information and professional advice I provide …* (CIEEM 2019).[10]

10 CIEEM (2019) *Code of Professional Conduct.* Chartered Institute of Ecology and Environmental Management, Winchester.

As this book is about writing reports, and because my experience comes from working as a consultant, I'm going to focus on the author's **impartiality** but clearly those tasked with reviewing a report must remain unbiased as well.

By stating an opinion you could, of course, be accused of taking a partisan view. And professional ecologists, by definition, are being paid for the service that they provide, so they always have a vested interest, to a certain extent. They must therefore work hard to remain unbiased, and to demonstrate their **impartiality**. In order to do so they should:

1. *Form opinions that are 'balanced'*
   In other words ecologists, when they write reports, should recognise that there may be an alternative viewpoint to their own.[11] They should give fair treatment to these different viewpoints. They are likely to need to come down on one side of the argument, otherwise the report may not be particularly useful, but they should discuss the opposite viewpoint, even if they disagree with it.

   It can sometimes be difficult to decide which alternative viewpoints to give 'air time' to, as it will be impossible to reflect on every possible different opinion. Here the report author will need to consider their target audience – what opinions might they have?

   For example, a widely used technique for moving water voles out of the way of the footprint of a construction project is 'displacement'. This involves manipulation of the habitat to encourage the animals to move of their own volition, rather than having to catch them and release them elsewhere. There have been a number of studies undertaken aimed at determining the effectiveness of this technique. Some have demonstrated that the animals don't move, whilst other studies appear to have shown water voles successfully moving out of the area affected by the works. Part of the problem is the difficulty of demonstrating success (whereas failure is often easier to demonstrate). And despite this it is still a recommended technique due to the need to deal with water voles in situations where the impacts would be so small as to make trapping the animals disproportionate to the risks.

---

11 By 'balanced' above, I mean consider both sides of an argument and come to a reasonable conclusion.

Any report needing to comment on the likely effectiveness of the displacement technique should therefore not claim it to be either categorically successful or unsuccessful – at least not at the time of writing this book. It doesn't necessarily need to go into a significant amount of detail about the various studies done, as many of these are covered by the relevant good practice guidelines (Dean *et al.* 2016),[12] provided that those guidelines are referred to, along with any more recent studies.

2. *Support an opinion*
   I've already discussed the need to provide supporting evidence for most opinions expressed (the exception being those that are widely held and uncontroversial). Where there are two sides to an argument, an **impartial** person will look at what the evidence suggests is the correct answer. Expressing an opinion that is supported by appropriate reference sources is unlikely to appear as biased (provided that it's done in the right way). Expressing an opinion that is unsupported by evidence, or where the evidence suggests that the contrary view is correct, could well appear biased.

   For instance, there is evidence to suggest that the behaviour of bats is affected by lighting (Stone 2013;[13] ILP and BCT 2018).[14] Different species may be more affected than others. Different types and levels of lighting will have greater or lesser effects. Given this, to argue the contrary position – that bats are not affected by lighting – would be difficult, as there is a significant body of evidence that bats are affected, and would need to be supported by credible evidence. However, an author could make an assessment based on common pipistrelles being less sensitive to light than lesser horseshoe bats whilst foraging, for example, although this would need to be supported with reference to evidence, such as set out in the sources listed above.

---

12 Dean, M., Strachan, R., Gow, D. and Andrews, R. (2016) *The Water Vole Mitigation Handbook (Mammal Society Mitigation Guidance Series)*. Eds Fiona Matthews and Paul Chanin. The Mammal Society, London.

13 Stone, E.L. (2013) *Bats and lighting: overview of current evidence and mitigation*. University of Bristol, Bristol.

14 Institution of Lighting Professionals and the Bat Conservation Trust (2018) Guidance Note 08/18. *Bats and artificial lighting in the UK. Bats and the built environment series*. ILP, Rugby.

3. *Try to be dispassionate*

   Most people that work in the ecology industry will be passionate about nature conservation – it's difficult to conceive of a situation where someone ended up with a career in ecology without caring too much about the natural world. And we can probably all agree that being passionate about nature conservation is a fundamentally important characteristic for an ecologist. However, it is difficult to be **impartial** if you hold a firm belief about something. And the more passionate you are about that belief the harder it will be to accept a counter-argument.

   Report authors therefore need to try to be dispassionate. This can sometimes be tricky to achieve. Perhaps the author of the report is a strong advocate of the conservation of a particular habitat or species, and the project is likely to have unfavourable outcomes for that habitat or species. The author needs to express those outcomes in a fair and balanced way, without letting their personal opinions cloud their professional judgement.

   For example, there are proposed development projects that, in order for them to be delivered effectively, will result in the loss of, or damage to, ancient woodland. A professional ecologist writing an EcIA Report for such a development will need to highlight the potential impacts on ancient woodland. They will need to assess the significance of the loss or damage based on ancient woodland being an important and irreplaceable habitat type. Dependent on where this project is located, there may be specific national and local planning policies stating that developments affecting ancient woodland should be refused planning consent unless there are exceptional circumstances. Whether or not those exceptional circumstances apply in this case is probably not a decision for an ecologist to make, as they normally relate to socio-economic issues – how important is this development?

   In this case, however much the report author believes in the protection of ancient woodland, and that the development is not sufficiently important to warrant an impact, they have to report the issues in a dispassionate way. Highlighting the facts without trying to lead the reader towards a particular viewpoint.

4. *Avoid emotive language*
   I've discussed this in an earlier chapter for reports in general – emotive language needs to be avoided. This is particularly important in allowing the author to demonstrate that they are **impartial**.

   For example, statements like this should be avoided:

   > *The proposed development is small-scale, comprising only 60 new houses, and therefore ...*

   The use of the word 'only' in this context, coupled with the term 'small-scale', implies something to the reader – that this proposed development is not really worth worrying about too much. In essence, it's not a big deal. However, some of the target audience may think that 60 new houses is quite a large-scale development in the specific context of where it is proposed, and may feel as though insufficient attention is being given to the issues as a result of its size being downplayed.

Proportionate and concise

The importance of being **proportionate** in what we write in a report, both in terms of how much we write (effectively, being as **concise** as possible) and in terms of any recommendations we make, considering both the degree of risk to biodiversity resources and the implications for the client, is (perhaps ironically) a big topic. I've therefore devoted the whole of the next chapter to it, rather than trying to cover it here.

Clear and precise

Clarity in the 'interpretation' sections of reports is vital. The reader needs to have a **clear** idea of what assessment is being made, or what is being recommended. This is where many authors seem to struggle, probably getting too bogged down in the rest of the report and forgetting that someone somewhere will need to take this report and 'action it'. In other words, you must remember your purpose and your target audience. Make sure that you communicate any recommendations or conclusions in a way that the intended audience will be able to identify and understand.

Precision is an important part of this. Given that someone will be undertaking some actions on the back of this report, you must ensure that they understand precisely what is required.

This means, amongst other things:

- Being careful about the way you phrase things;
- Doing your best to minimise the chances of misinterpretation;
- Avoiding vague statements;
- Structuring the report to ensure that any key conclusions or recommendations are obvious.

Here are some examples:

1. *Recommendations or requirements*
   A Preliminary Ecological Appraisal (PEA) might identify an ecological feature as being a key constraint to a proposed development – let's say it's a Local Wildlife Site within the site boundaries. The ecologist writing the PEA Report might consider it unlikely that the development will be considered acceptable by the Local Planning Authority as it is currently proposed, due to the impact that it would have on this Local Wildlife Site. The ecologist may also have identified another ecological feature of less significance – let's say that there's a small patch of scrub used by a range of common bird species for nesting; this will have implications for the construction programme and protection measures, to ensure compliance with the legislation relating to nesting birds. Both issues need to be raised in the *Results* section of the PEA Report and the implications of them, and any recommendations arising from them, discussed in the 'interpretation' sections.

   In this case it will be important to separate out the two issues clearly in the report – ideally having them in separate subsections with appropriate headings, such as 'Design constraints' and 'Protection measures required during construction' for example. The text relating to them also needs to make the reader immediately aware of the difference in terms of the implications of these issues.

The text might be something along the lines of what I've set out below.

Blackberry Wood is designated as a Local Wildlife Site. Policy E3 of the Local Plan (see Section 2.2 of this report) states that 'developments likely to have an adverse effect on Local Wildlife Sites should not normally be permitted except in circumstances where the benefit of the development outweighs any damage caused'. The development, as currently proposed, would result in the loss of the Local Wildlife Site. It is difficult to weigh this loss against the benefits of the development, but this should be considered a major risk to the project, as it may be considered to not accord with this planning policy. It is therefore recommended that the proposed development is redesigned to …

The patch of scrub identified by Target Note 6 on Figure 2 is used by blackbirds and wrens, and possibly also by other common bird species, for nesting. All wild birds are protected whilst nesting under the Wildlife and Countryside Act 1981 (as amended). Measures to ensure compliance with the legislation will be required during site clearance and construction. Therefore vegetation clearance should be timed to take place outside of the period when birds are likely to be nesting (i.e. should not take place during the period XXX to XXX inclusive) or should be preceded by a survey confirming the absence of nests.[15]

2. Specific details of recommended actions
A proposed development will result in the demolition of buildings used by birds for nesting. As compensation for this loss it is proposed to install nest boxes on retained trees and to provide integrated nest boxes within some new buildings.[16] This is set out in the EcIA Report accompanying the application. It is important that the report provides specific details of the new nest boxes:

---

15 I've deliberately not specified the months that start and end the bird nesting period. Clearly this differs for different species and in different parts of the world. However, note the way that the text in brackets is phrased. If I'd simply inserted the months marking the start and end of the nesting season, whatever I'd considered those to be, then the text could have been misinterpreted as the dates when vegetation clearance should take place, rather than the dates when it should not.

16 It is worth mentioning here that, at the time of writing, a new British Standard is being developed on specifications for the design and installation of integrated bird nest boxes.

- How many are proposed?
- What sorts of box are proposed?

  (You might not want to be too specific here as a certain box might suddenly become unavailable, but should specify the target bird species that the box is designed for, and the materials that it must be constructed from, as a minimum)

- Where are they proposed?
- Is any maintenance or monitoring proposed?

Reports without 'interpretation' sections

Occasionally a professional ecological report might be written that does not have any of those 'interpretation' sections. Given the importance of interpreting findings in a report, this is only likely to be the case where a report has specifically been written to exclude those sections. For example:

- A report intended to be an appendix to another, and therefore containing only survey methods and results;
- A Landscape and Ecological Management Plan or Habitat Management Plan, which has been written to set out management objectives and prescriptions; or
- A report produced by a subcontractor for another professional ecologist, where the scope of work of the subcontractor was simply to set out the methods and results of the surveys they undertook, leaving the ecologist who contracted them to provide the interpretation.

This will only be relevant in a small number of cases. For the most part a client commissioning an ecological study is going to be much less interested in what's been done and what was found than they are in what the findings mean in the context of their project. Similarly, other likely target audiences, such as Local Planning Authorities, nature conservation consultees, or people living in the local area around a proposed development project, will want to know what the implications of a project are for biodiversity.

Writing the Conclusions section

Many professional ecological reports will need a *Conclusions* section. There are some clear exceptions, like a Habitat Management Plan or a Biodiversity Strategy – both documents set out a series of actions to be undertaken as well as the basis for prescribing those actions, and therefore don't need to draw a specific conclusion. However, ecological survey reports, PEA Reports, EcIA Reports, Research Reports, Reports produced as part of the Habitats Regulations Assessment process, Biodiversity Net Gain Reports, and Ecological Monitoring Reports, for example, will all need a *Conclusions* section.

The *Conclusions* section is important. Remember that journey I talked about in Chapter 1? Taking the reader through a series of logical steps from the initial purpose or purposes to an overall conclusion (or conclusions) – and the conclusions have to relate directly back to the purposes. The reader is therefore looking for a *Conclusions* section in the reports I've listed above (as well as several others). If you don't write one you aren't really fulfilling the purpose – you're providing some information and letting the reader draw their own conclusion, and that's not normally the point of a professional ecological report.

Writing the *Conclusions* section then should be pretty straightforward once you've completed the rest of the report (apart from the *Summary*) provided you've written a clear purpose. If the purpose is clear, and the report has taken a series of logical steps towards achieving that purpose, the conclusions should be obvious. If they aren't obvious, it's probably because one of these two things has gone wrong.

Importantly, the *Conclusions* are not the same as a *Summary*. The *Summary* needs to include a brief explanation of the purpose, what's been done, what's been found, and what this means (see Chapter 13 for a detailed explanation of how to write a *Summary*). The *Conclusions* need to be clear statements that relate directly to the purpose or purposes. These statements must be supported by the text in the report.

Try to keep the *Conclusions* short and focused. Think about the key messages that your target audience needs to obtain from your report and stick to those.

Chapter 10

Keeping it in proportion

One of my Key Characteristics of an effective ecological report is being **proportionate and concise**. The definition of concise in the New Collins Concise Dictionary of the English Language (and, let's face it, they should know) is given as '*brief and to the point*'.[1] The Cambridge Dictionary defines it as '*short and clear, expressing what needs to be said without unnecessary words*'.[2] I prefer this latter definition, as it highlights the importance of saying '*what needs to be said*'.

This Key Characteristic needs to be delivered on at least three different levels:

1. The length of the report;
2. The relative amount of discussion of individual topics within the report; and
3. The costs versus benefits of recommendations or commitments included within the report.

---

1 New Collins Concise Dictionary of the English Language (London and Glasgow: Collins, 1982).

2 https://dictionary.cambridge.org (accessed 2 November 2020).

Length of the report

Let's start with the first of those aspects. Reports should be as short as possible. They need to communicate a specific message, as set out in the *Introduction* as the purpose(s). Anything that isn't relevant to that purpose (or purposes) is likely to detract from communication of the message.

So, we need to make sure that our reports give a fair treatment of the issues – they shouldn't be short for the sake of it. This, effectively, means being **proportionate.** A report for a complex site with numerous valuable habitats, or with major issues that require detailed discussion, should be longer than one for a straightforward site or with relatively minor issues. When we pick up an ecological report and think to ourselves 'that looks like a long report' we are expecting to find that there's a lot of important information included, and that the specifics of the site or the project dictated the need for the weighty report that was produced. This, however, isn't always the case.

I was asked to review an EcIA Report recently. It was for a very small site, less than a hectare in size, in an urban location with no designated nature conservation sites and no valuable or important habitat present. It did have a couple of protected species issues to assess, and there was an invasive non-native plant species present, requiring an explanation of the legislation relating to it and appropriate control measures. Nevertheless, I was amazed to receive a report that was nearly 70 pages long.

Why was that report so long? Did it contain lots of useful information? Well, unfortunately, the answer is not enough to warrant 70 pages. I think I could have done the same job in 20 pages at a push, 30 pages at absolute most. The excessive length was, as far as I could see, due to two specific reasons:

1. There was far more explanation of EcIA process than was required; and

2. There was a considerable amount of standard text that probably appears in every report produced by the same company, which didn't really add anything and which nobody will bother to read.

Now, don't get me wrong. I like to see an explanation of the EcIA process in an EcIA Report – I think that's important. But that shouldn't outweigh the amount of text devoted to describing the key outcomes. Who, after all, is going to read it? And if they read it, will they understand it? The process could have been described far more simply in this particular case.

And many ecologists will include things in reports to counteract the fear of leaving out something important – better to have too much in than to miss a key piece of information. This is understandable, up to a point. However, the author should work out what information is relevant to their target audience and include that, rather than sticking in a load of stuff, some of which may be relevant but some won't, without any real thought.

In the case I described above I think both of these problems originated from the fact that the report was produced following a standard 'report template' that had been developed over the years to fit many different scenarios including (presumably) some large and complex sites. The outcome, when it was applied in this particular situation, was overkill – far more information than was needed to do the job. Report templates are discussed further in Chapter 12.

Why is this important? Does it matter if a report is longer than it needs to be? Well, it certainly isn't the worst mistake that a report author can make. And deciding whether a report is too long or not is a subjective opinion – others might have read a 20–30 page report that I produced for the same site and considered it to be too short.

The problem with reports that contain unnecessary information is that the valuable and important bits can become hidden in amongst the superfluous. It can detract from the report's central aim of communicating its message to the target audience, in relation to the specific purpose(s) set out in the *Introduction*. The reader may skip bits because they can't be bothered to read stuff that doesn't matter to them, and as a result may miss information that is vital to their understanding.

We must, therefore, minimise the amount of 'standard' text that's included, such as a description of the background biology of a species, or the detail of the legislation protecting it. It might seem OK to provide this as an appendix at the back of the report – at least then it won't be so distracting to those members of the target audience that don't need to read it. However, if it's relevant to the purpose of the report then it should be in the main text. If it's not relevant then why is it in there at all? I've read reports where there's almost as much 'standard' text in the appendix as there is in the rest of the report – if it was cut out the report would be about half the length.

Perhaps professional ecologists think that clients expect to see a big meaty report for their money. Well, maybe some do! My experience, though, is that clients want a report that is readable – one that they can understand – as well as one that fulfils the purpose required of it. If it can achieve this whilst being short, so much the better.

Here are a few tips then for keeping a report as short as possible:

- Get the *Contents Page* onto a single page (see Chapter 7).
- Minimise the amount of text describing legislation – only include it in detail if it is relevant.
- Don't include 'standard' text in appendices. If it's necessary to understand the report – include the information in the main body of the report. If it isn't – delete it.
- Avoid repetition wherever possible (I'll probably mention that again!).
- Think twice before including detailed descriptions of standardised methods where there are no site-specific factors to be discussed. For instance, I may have taken a sample of water from a pond on a site and sent this off to a lab for analysis to see if it contains environmental DNA from great crested newts. If I've collected the sample using a standard protocol, and the lab have analysed it following a standard protocol also, a reference to those protocols is probably all the detail I need in my report.
- Don't say the same thing twice, just in a different way (I suspect I've probably said that already!).
- Keep your focus on the purpose of the report. If you're writing something that isn't directly relevant then consider whether you need it at all. If you decide that you do, then give it very brief treatment.

Relative amount of discussion of individual topics

You should, of course, apply the principle of keeping the overall report **concise** to each individual section as well. However, a good report will also find an appropriate balance in terms of the amount of text in each of the sections and subsections, devoting more discussion to key and complex issues, and less to the straightforward ones.

There's a tendency, particularly when first starting to write professional ecological reports, to provide more information, and greater levels of detail, in relation to the topics we know about or have a particular concern for. This is obviously something we need to try to avoid.

Every time we sit down to write a report we need to think carefully about our target audience – which sections are they likely to be particularly interested in, and which are they likely to skip over fairly quickly? The factual sections of an ecological report, the *Methods* and *Results*, are normally the easiest ones to write. Yet these sections, whilst important, are often the ones that fewest members of the target audience will be that interested in. Most of the target audience will want to know what the outcomes of the surveys mean in the context of a particular project, for example.

EcIA Reports often have very long *Methods* sections. This is because the report will be underpinned by numerous different surveys and the author will, of course, want to make sure that the description of the methods is **robust** and provides sufficient detail. And the detail does need to be included in the report. However, a lot of the factual stuff can be provided in appendices to keep the text in the *Methods* section as short as possible. Tables of data, dates, times and weather conditions of surveys, standardised protocols for surveys, can all go into the appendix. What should be in the main text is a summary of what was done, any explanation of why it was done and why a particular approach was taken, such as accordance with or departures from good practice guidelines, and any limitations.

Similarly the *Results* sections of some reports can be made more **concise** by moving the hard facts to an appendix. This should then allow a greater proportion of text in the main report to deal with the implications of the results, rather than necessarily the results themselves.

Within each of the sections dealing with results, or interpretation of those results, it is also important that the most pressing issues (as far as the target audience is concerned) get a greater level of treatment than those of lesser relevance. For example, take a Preliminary Ecological Appraisal Report that identifies a series of ecological constraints to a development proposal: one that might be a 'show stopper' as far as that development is concerned; one that won't stop the development but will probably need to be designed around; and one that doesn't need to influence the design and can be dealt with through simple on-site protection or mitigation measures. The first of these constraints, the potential 'show stopper', will likely warrant the greatest level of explanation. It should also be discussed first. The second should be addressed next, and will also need a reasonable amount of discussion, but probably less than the first. The third, the constraint that doesn't need to influence the design and can be dealt with through simple on-site protection or mitigation measures, will need relatively little text in comparison with the other two, and should be discussed last.

Costs of recommendations or commitments versus degree of risk (or likely benefits to) biodiversity

The final aspect of being **proportionate** that I want to highlight relates to the recommendations or commitments being made in an ecological report. When we write reports we need to remember that others are looking to us, as professional ecologists, to give expert advice. They may put a considerable amount of faith in what they read. If we say that a certain action should happen or really shouldn't happen then this can be taken 'as gospel' by some readers.

This is an important lesson for professional ecologists to learn. When an ecologist starts their career they might be told to write a report with very little guidance, and so off they go and do their best to cobble something together. What they won't necessarily realise is that the specific words they have written, the precise way they have phrased something, and the particular emphasis they have given, will make a difference to how that report is interpreted, and this is likely to have repercussions.

With experience that ecologist will learn that they need to craft their report carefully. They will need to 'guide' the reader appropriately, allowing them to understand the issues, without leading them to a certain viewpoint. And they will need to think carefully about the recommendations they make, and the implications of any commitments they make on behalf of their client.

When we write recommendations or commitments into a report then, we need to ensure that we are **proportionate** in what we propose. Here are a few examples:

1. An ecologist has been commissioned to carry out mammal surveys at a site, comprising a small area of farmland. The first stage of this involves a desk study that identifies records of polecats within 2km of the site. The site provides suitable habitat for this species (although so does the vast majority of the land within the surrounding area). Given this an ecologist might recommend a field survey for polecats to confirm their presence or likely absence. There are no straightforward survey techniques that can be used for polecats – options that have been considered are camera trapping over a large area and searches for scats to be sent off for DNA analysis. Both options would be expensive. Would a survey be **proportionate** in this case? Could the presence of polecats be assumed and an assessment made without a detailed survey?

2. An ecologist has been commissioned to undertake a Preliminary Ecological Appraisal of a small site proposed for development. The site is arable land with an area of woodland directly adjacent to one of the site's boundaries. The ecologist recommends retention of a buffer zone of 30m alongside the woodland but this would result in half of the site being undevelopable. Is this a **proportionate** recommendation? Is the woodland of sufficient importance, and the potential impacts on it so great, as to warrant such a buffer? Is there any guidance or policy that suggests such a buffer zone is required? Or has the distance of 30m been plucked out of thin air? Maybe it's possible that a buffer zone of 30m is appropriate in a particular case and can be justified. But it would need to be thought through carefully, given its impact on the proposals.

3. An ecologist has been commissioned to undertake an Ecological Impact Assessment for a proposed development, which needs to include mitigation measures to be committed to by the developer. A length of hedgerow needs to be removed to allow construction of site access. There is a chance that birds will nest in the hedgerow and hazel dormice are known to be present and may hibernate in the base of the hedgerow.

The ideal timing of vegetation clearance to avoid (or at least minimise the risk of) damaging nests or disturbing nesting birds would be something like September to February inclusive (although there will be seasonal variation to this). The ideal timing of the removal of habitat to minimise the risks to dormice might be very different, as dormice hibernate during winter and could be at risk if works are undertaken during the period October to April. The ecologist will need to weigh these two risks and will need to consider whether they are so great as to warrant restricting the developer to one or two specific months when they can remove the hedgerows. Or whether an alternative approach can be considered – perhaps a two-stage approach to hedgerow removal to balance the risks to both birds and dormice, for example.[3]

4. A proposed development will result in the infilling of a 20m length of ditch that supports water voles at relatively low densities. The ecologist advising the developer recommends compensation for this loss in the form of new wetland habitat. This seems like a sensible approach, but the key question is how much would be appropriate? If the developer happens to be creating wetland habitat within the site in any case, and can amend its design to make it valuable for water voles then this may not be a particularly onerous proposition. However, if no wetland habitat is included as part of the project, and the developer would need to buy additional land to deliver this, the amount of habitat needed will be fundamentally important. In either case it will be important to assess the impact of fragmentation associated with the infilling, and whether the proposed location for any compensatory habitat is sufficiently well linked to the locations where water voles are present. The ecologist will need to be well informed as to what is normally considered appropriate in such circumstances to allow them to provide **proportionate** advice.

3 Note that hazel dormice are protected and therefore a licence from the relevant Statutory Nature Conservation Body would be required if there was a likelihood of an offence being committed, and the appropriate approach would need to be agreed with them, through the licensing process.

Chapter 11

Tables, figures, photos and appendices

In the first chapter I highlighted the importance of clarity when presenting information (my Key Characteristic Number 10 was that reports should be **clear and precise**). This is vital to ensure that reports are accurate and sufficiently detailed, whilst still catering to the target audience. In this chapter I'm going to discuss the use of various means of presenting information beyond simple written text in the main body of the report: tables; figures; appendices; and photos.

In general terms these means of presentation will be used for one of three specific tasks:

1. To **effectively present** large quantities of data.

   Tables are often used for this purpose. Graphs may also be used in some cases.

2. To help the reader **navigate through** the information presented, allowing them to easily find their way to the information they want (and thereby also helping them skim read, or avoid completely, the information they don't want).

   At the basic level this might mean providing a *Contents Page*. Appendices and summary tables also fall into this category and are particularly helpful when writing reports where the target audience covers a broad range of perspectives and knowledge.

3. To help the reader **understand** the information presented through the use of appropriate graphics, bearing in mind the expected level of knowledge of the target audience in relation to the site, the project or other technical details.

   Figures or drawings and photos fall into this category.

Contents Pages

Let's start with, hopefully, the most straightforward example. A *Contents Page* is effectively a summary table listing the headings and perhaps also sub-headings (as well as sub-sub-headings in some cases) and the page number where each commences. This allows the reader to navigate through the report. It also allows them to get a brief first impression of what's going to be included.

I've already discussed *Contents Pages* in detail, in Chapter 7, so won't spend any more time on them here.

Tables

Tables can be a very helpful way of presenting lots of data – data that would be difficult to read or understand in plain text. In the context of an ecological report, this might include:

1. Tables providing details of survey methods. Many surveys will require multiple visits to a site, and the dates, times, weather conditions and personnel involved may vary for each visit. In many cases these details will need to be provided. A table is the best way of presenting this sort of information.

2. Tables providing details of survey results, such as the number, sex and life stage of slow-worms recorded on each reptile survey visit, or the number of passes by each different species of bat recorded on an automated bat detector during each night of survey.

Tables are also a good way of summarising information, either to summarise something that has already been presented in text, perhaps appearing at the end of a particular section of a report, or instead of the text, in cases where a table presents the information more clearly. For example:

1. A Preliminary Ecological Appraisal (PEA) Report may include a significant number of recommendations, in terms of design changes to avoid or minimise ecological effects, general mitigation measures likely to be required, further surveys to inform any subsequent ecological impact assessment, and possible enhancement measures. These will need to be discussed in the text, but a summary table for each could be very helpful for the target audience to pick out the information that they specifically require.

2. Ecological Impact Assessment (EcIA) Reports will include mitigation measures. These will be described in the text of the report, and these descriptions could be scattered throughout the *Assessment of Effects and Mitigation Measures* section. It will be helpful to the target audience if a table is provided, summarising all of the mitigation measures proposed at the end of this section.

3. In a dormouse survey report, an assessment of the various hedgerows on a site, in terms of their suitability for dormice, could be confusing if written in plain text. A table, with a reference number for each hedgerow, a description of the various criteria that might influence a hedgerow's suitability for dormice (age of hedge, height, width, species composition, frequency and type of management, etc.), and an overall assessment of suitability would be easier to understand. This should be cross-referenced to a suitable figure to allow the reader to understand the location of each hedgerow. Which leads me nicely on to …

Figures, drawings, maps or plans

Figures, drawings, maps or plans (I'm going to refer to them as figures in a generic sense from here on) are an essential way of presenting information that is location specific. And because the location of specific features, or the extent of certain habitats, will be key issues to communicate to the reader in most ecological reports, virtually every ecology report will need at least one, if not two, figures. In some cases, several more will be needed.

Let's start with the basics. Most ecology reports will relate to a specific site, or area of land. The location of that site will be essential information to the reader – not just so that they understand where it is, but also so that they understand the context of the general landscape that it is located within. The first figure to include will therefore be a 'site location plan'. These should:

1. Be presented on an appropriate base map – either a suitably scaled[1] Ordnance Survey map (or their equivalents) or an aerial photo;

2. Include a clear site boundary;

3. Include sufficient area around the site to provide context – a site location plan that gives little detail beyond the site boundary will be difficult to interpret; and

4. Include labels, as necessary, to pick out key features.

In relation to items (3) and (4), imagine you are trying to find the site for yourself from the site location plan – is there sufficient information provided to do so? And if you are describing the site's context in words, are the features described visible on the site location plan and appropriately labelled?

The other figures to be included in a report will depend on the type of report and its complexity. Normally a 'site boundary plan' will be needed. Note though that a 'site boundary plan' and a 'site location plan' are not the same thing. The 'site location plan' will need to show the context of a site over a sufficiently wide area. The 'site boundary plan' will normally need to be presented at a greater degree of resolution, which will mean losing the surrounding context.

The study area for a survey will normally need to be provided on a figure, unless the study area is the same as the site (in which case the 'site boundary plan' will suffice).

Where a survey involved the walking of transect routes, or the use of vantage points (for birds) or listening points (for bats), these will need to be shown on a figure (titled 'study area' or 'survey methods' or something similar).

---

1 The scale will depend on the size of the site, but 1:25,000 scale Ordnance Survey maps are appropriate for many sites.

Without this information it is likely to be difficult to assess the validity of the surveys, and it will not be possible for another surveyor to repeat the surveys and get comparable results (the benchmark for determining how much detail to include on survey methods – see Chapter 8).

Surveys that only took place in specific locations within a site will need a figure showing these locations. For example, if certain parts of a site were deemed to be suitable for use by reptiles and were the subject of a targeted survey, a figure identifying those areas will need to be included. If several different ponds were the subject of an assessment in terms of their suitability for great crested newts, these should be labelled individually and their specific locations shown on a figure.

Survey results will almost always need to be provided in the form of a figure, as well as discussed in the text. The text and the figure should be consistent with each other, so that the reader can understand the text better by looking at the figure, and vice versa. The exception to this will be if the survey didn't record anything, or if there's nothing to add beyond what's shown on the 'study area' or 'survey methods' figure.

A PEA Report, for example, is likely to need a figure showing the habitats present within the site (and possibly extending off site) and a figure showing the location of designated nature conservation sites relative to the site (given that the search area for designated sites normally extends some distance beyond the site, this information will normally be provided on a plan that has been zoomed out from the site).

A figure won't always be needed to show the results of an ecological survey, but it can often be very helpful, depending on the complexity of those results. For example, a bat activity survey might well have recorded significant amounts of foraging activity by different species in different parts of a site, and perhaps a commuting route used by one particular species to move to and from a roost site. This information will almost certainly need to be shown on a figure as it will be too complex to describe accurately to a reader in text only. You could argue that every single bat pass should be shown on a figure in some cases, rather than a summary of the main areas of activity or key features – it will depend on the purpose of, target audience for, and likely uses of the report.

All figures will need an appropriate base map – normally a suitably scaled Ordnance Survey map (or their equivalents) or an aerial photo.[2] A figure showing the habitats present on site, as well as often being needed in its own right, will also provide a useful base for some of the other figures needed. For example, the bat survey transect route will undoubtedly have been designed to sample a range of habitats and different parts of the site, and overlaying it on top of the map of the habitats present can therefore be very useful.

In some cases it's possible to combine multiple topics on the same figure. This can help to reduce the length of a report. However, it only works if all topics require the same base map and work at the same scale. It's also important to avoid excessively cluttering a single figure with too many pieces of information, as the value of the figure will be lost. And information that is inherently linked will work well on the same figure, whereas unrelated information doesn't work so well. For example, if there was sufficient space, you could produce a single figure showing the bat survey transect routes and the key survey results. It probably wouldn't work that well if you included the areas surveyed for reptiles on the same figure.

And in other cases you might need multiple separate figures to illustrate one single 'topic'. This will be the case where there's a lot of information to display, or the site is linear, meaning that it won't fit well and at a helpful scale on a single page. You need to consider the reader here. When they print the report off (some will want to read a hard copy) will the graphics be legible? Or if you reduce the figure to A4 size will it be impossible to interpret? And if you include an A1 size figure in amongst an A4 report, this might be OK when reading an electronic version where you can zoom in or pan out, but when it comes to printing it this is likely to be problematic.

Importantly, for ecological surveys undertaken in relation to a specific site, the site boundary will need to be shown on each figure as a reference point.

Each figure will need a key, so that any symbols used are clearly explained.

Each figure should also include an indication of scale. However, the accuracy of the figure will be dependent on the accuracy that the information could

2 The scale of the base map needs to be sufficient to show relevant features. It is difficult to be prescriptive about the appropriate scale of mapping as this will vary in different cases. However, 1:10,000 scale Ordnance Survey maps are suitable in many cases.

be plotted at. In some cases an approximate location for an ecological feature is sufficient, but this needs to be made **clear** to prevent others trying to scale a specific location off the figure if that wasn't intended.

Each figure should also include an indication of orientation – a north arrow is the simplest way of doing this.

Where a set of multiple figures is provided within a report, perhaps illustrating the survey methods and results from each of a range of surveys, these should try to use a consistent base map to aid interpretation. Several years ago the company I worked for was awarded a project for which many of the baseline ecology surveys had been completed by another consultancy. The figures showing the badger survey results were at a different scale, and a different orientation, to the figures showing the habitat survey information. This made life unnecessarily complicated and it took a while to work out that the badger survey and habitat survey actually had significantly different study areas due to a change in the design of the scheme between the two surveys taking place.

The orientation of the figures should therefore be consistent throughout. What I mean here is that north should always be in the same direction on each figure rather than all the figures should be portrait or landscape. That can help as well, but sometimes you won't be able to avoid switching between portrait and landscape orientations. Incidentally, it helps if figures are orientated with north at the top of the figure for reasons of convention. There might well be a reason to orientate differently in some scenarios, but north should normally be 'up', as a general rule.

Confidentiality and copyright are important issues for reports in general. In relation to figures there are a couple of issues to consider:

1. Who has the ownership of the base map? For example, if you're using an Ordnance Survey base map you will need an Ordnance Survey licence, and will need to follow the conditions of that licence (which will include copyright information that needs to be quoted on each figure).

2. Who has ownership of the data? Is it yours, or are you using a figure provided by another person or organisation, for which you need permission and an appropriate acknowledgement.

3. Does some of the information being shown need to be kept confidential? For example, it might not be appropriate to show the specific location of badger setts in a document that will be publicly available. However, the location of the setts is likely to be important to an understanding of the issues relating to badgers (dependent on the purpose of the report). This is likely to be most easily resolved by including the badger information on a separate figure from any other information and clearly labelling it as confidential, to be redacted from public versions of the report.

Finally, what doesn't need to be included on a figure? Well, the short answer is anything that is:

1. *Not relevant to the aims of the report*
   Incidental records of something that wasn't the subject of the survey might come into this category, although in some cases it might still be worth including these.

2. *Not location specific*
   For example, the fact that a brown hare was spotted during a walkover survey of arable farmland for a PEA Report is of relevance, but the precise location where it was spotted is probably not. In this case it should be mentioned in the report but doesn't need to be shown on a figure.

Photos

Photos are invaluable in ecological reports. When I first started writing professional reports you wouldn't dream of including a photo. And it wasn't all that easy either, because you'd have to get your film developed, the photos printed, and then the photos scanned in and then manipulated into the report. This was, invariably, far too much hassle. Not to mention the hassle of carrying a camera with you on a site visit along with your binoculars and clipboard and field guides – just another thing to lug around and probably lose!

Things have changed. Pretty much everyone now carries a camera with them at all times, in their phone. And the photos can be uploaded, sorted through, and copied into a report in a very short space of time. It's now more usual to see professional ecological reports with photos than without. And this is a very good thing. There are a few issues to consider here:

1. What sorts of photo should be included?

2. Where should photos appear in the report?

3. How should photos be labelled?

In general terms, I'd say that a photo should be included if it adds value to the report. This might be because it illustrates the text, allowing a reader to piece together lots of information in the correct manner. For example, in a report I might be describing the suitability of a watercourse to support water voles, with reference to the width and flow rate of the watercourse itself, the bank profile and substrate, the type, density and extent of emergent vegetation, the amount of shading, and maybe some other factors as well. A well taken photo will assist the reader in understanding the description.

Photos that show the general context of a site can also be very helpful – remember that the vast majority of people who read your report won't have visited the site.

In some cases a photo will be extremely helpful in recording a specific feature or location. For example, in assessing the suitability of a building to be used by roosting bats I may have identified several features as potential access points or possible roosting features. Being able to identify the specific location of these features at some point in the future is likely to be important – perhaps for more detailed further surveys, or during the demolition or conversion of the building. Having a report with well taken photos included, with sufficient context to allow those features to be identified, can be invaluable. The key here is to make sure that the photo has sufficient context in it to allow the location to be identified – a close-up photo of a hole in the ground as a possible otter holt will probably not be as helpful as one taken showing some of the nearby features.

In some cases I've seen photos included in a report that don't really add value. Multiple photos of badger dung pits, for example. Now I'm as obsessed by animal excrement as the next ecologist, but will it help the reader to see a close-up photo of a hole in the ground with some sloppy poo in it? I'm not convinced it will. Which member of the target audience, upon reading the words *'a badger dung pit was recorded (see Photo 6 in Appendix 2)'* in a report, will find themselves eagerly flicking to Appendix 2 to get a good look at some badger poo?

That's not to say that you shouldn't take the photos when out in the field – perhaps to remind yourself of what you found, or to show to a colleague as 'evidence' or to confirm your identification. But that doesn't mean they all have to go into the report – perhaps include one 'typical' photo to demonstrate that you've identified something correctly.[3]

There are also cases where a photo of animal droppings can be very important to include, beyond confirming 'identification' – such as a photo showing the number and distribution of bat droppings beneath a roosting feature.

Some ecological monitoring will involve the use of fixed point monitoring to detect changes over time. The inclusion of the photos in the report will clearly be vital in such cases.

Assuming then that we've taken some useful photos to include in our report, where should they go? There are two basic options. Firstly, you could include them within the main text, at the location where the point they are illustrating is being discussed. This is really helpful for photos that fall into the category of illustrating a point in the text. It's less necessary for those photos that might allow a specific feature to be identified at a later date – I'd suggest that such photos would probably sit better in an appendix to the report. And having photos in the main body of the text only really works well if there's no more than one photo every few pages. When you need to include multiple photos to illustrate points discussed in close proximity in the report it will be better to move them all to an appendix.

Photos should be numbered in a logical manner throughout the report. This could be as simple as in the order they appear: Photo 1, Photo 2, Photo 3, etc. Or it could include the section of the report they appear in: Photo 1.1 (which would be the first photo in Section 1), Photo 2.1 (the first photo in Section 2), Photo 2.2 (the second photo in Section 2), etc. And there are probably other

---

3 I recently undertook a survey for a particular protected species, using an appropriate search technique and at the correct time of year, but failed to find any of its characteristic field signs. This was surprising as a different ecologist had recorded it as abundant in the same location a few years earlier, having recorded numerous field signs. There are several possible reasons for the disparity, one of which became obvious when I reviewed their report. The report did contain photos of the field signs and I could clearly see from these that they had incorrectly identified at least some of the field signs (the ones they provided photos of). Of course the species in question may have been present and been lost from the site for other possible reasons, but the photos were very useful in this case.

ways of doing it that will work equally well. Reference to the original image number is generally not going to be helpful though: IMG0239, IMG4517, etc. – this will be very confusing for the reader.

It's also helpful to give each photo a caption, such as 'Photo 1: general view across site from the southern boundary' or 'Photo 2: Pond A'.

Finally, it's worth making sure when you take photos in the field that they:

1. Show what you want them to show;
2. Are in focus; and
3. Are of sufficiently high resolution to allow them to be zoomed into without becoming blurry.

You should obviously do these checks in the field. There are no excuses for blurry photos in a digital world!

Other illustrations or drawings

In some cases it might be useful to include other ways of communicating information that is difficult to describe in words, such as an illustrative drawing to show the proposed profile of a new pond being created, or the idealised design of a reptile hibernation site.

Appendices

When appendices are used they are provided at the end of a report. This is to allow sufficiently detailed information to be included in the report, to allow it to meet its objectives, whilst minimising any interruption of the flow of the text in the main body of the report. Multiple tables of data, such as those referred to above, can be included in an appendix. The information contained within them can then be referenced in the main report and perhaps summarised.

The appendices could also include details about a specific survey technique or mitigation measure. For example, if an artificial otter holt is proposed or recommended in a report then a typical design of such a structure could be useful – this could be included in an appendix. Similarly, in order to fully describe a survey technique it may be necessary to include some typical details, such as an explanation of how torchlight surveys for amphibians are undertaken, or how bottle traps are constructed and set (again, for amphibians). This information could be provided in the main body of the report, in the *Methods* section, but this may cause the report to become unbalanced in terms of the description of survey methods for amphibians in comparison with other species groups, or it may be felt that this extra detail makes the *Methods* section very long and most of the target audience won't bother to read it. The alternative, then, is to move this detail to an appendix, suitably cross-referenced in the main text.

Appendices can also be useful for providing photos, as discussed earlier in this chapter.

My general rule for deciding what can go into an appendix is that it should be 'facts' only, such as survey methods, survey results, or typical details. It shouldn't include assessment or interpretation, such as an explanation of why a survey has been designed in a certain way, or an explanation of the survey limitations.

Something I regularly see in ecology reports is the use of appendices for standardised text that appears in pretty much every report that's produced. This might be a description of the background ecology of a species, or its specific legal protection, for example. The problem with this is that a report often ends up containing a significant amount of text that isn't relevant as it's included in all reports as standard. If an element of a species' background biology, or their habitat preferences, or their legal protection is relevant to the aims of the report, then this information should normally be provided in the main text, appropriately referenced.

In some cases it might well be appropriate to include the specific wording of legal protection in an appendix, particularly where it applies to several different species referred to in the report. Putting it into an appendix will reduce the amount of repetition. The specific planning policies of relevance to an EcIA Report or a PEA Report could also be reproduced in an appendix, cross-referred to in the main text as needed.

The appendices of a report, then, should be as 'bespoke' as the main body of the report, and not used for standardised text that may or may not be of relevance.

This is one of the inevitable problems that arises from having a standard report 'template', which I'll discuss further in the next chapter.

Chapter 12

Creating and using a template

Many companies or organisations will have their own report template. Effectively, this is a starting point for a report, including *Title Page*, section headings and sub-headings. It will normally also include some key pieces of text – those that are standard across all reports. The template might also include some text that may or may not be relevant, dependent on the scenario, which the author needs to use, edit or delete. And it will, of course, be set up using a 'house style', in terms of format, layout, numbering, font type and size, headings, and perhaps some other details.

Why use a template?

There are some clear advantages to using a template, but also some drawbacks. These are worth considering if you are thinking about creating a template. They're also worth considering if you are already using a template, as knowing the obvious pitfalls can help you avoid them.

There are two main advantages of using a template. The first of these is consistency. This includes a number of different aspects, each with their own advantages:

1. The author is prompted by the template to include key pieces of information

   This is a major benefit of a template. There is specific information that must be included in a given report, or that a particular company or organisation may have decided should always be included. Having prompts in the template that remind the author to include these will mean that they are less likely to be forgotten about.

2. The style, layout and content of all reports produced by a company or organisation are consistent

   Many companies or organisations will value consistency in this regard particularly highly.

3. There is an agreed way of phrasing a particular issue or set of circumstances

   Sometimes a template will have been written in such a way that it provides a standard piece of text that fits, pretty much, with the scenario that an author might be facing for a given report. This will help guide the author to deal with that issue in a manner that is consistent with how it has been treated in the past, how it will be treated in the next report (assuming nothing changes the approach), and how it would be treated by any other ecologist working for the same company or organisation.

   This will be very helpful for junior members of staff, or those with limited experience of report writing. It will obviously also be beneficial to those tasked with reviewing or editing reports, as they shouldn't have to keep amending the same 'mistake'.

These aspects of ensuring consistency will be beneficial for anyone writing multiple reports. However, the amount of benefit will depend, to a certain extent, on the size of the company or organisation. The consistency provided by a template will be hugely advantageous for large companies or organisations, as it would be very difficult for them to ensure this in any other way. An individual ecologist, working for him or herself, may not see any real advantage in having a template. There are other ways of getting some of the advantages, which I'll come on to shortly, and there are some major disadvantages, which I'll also discuss.

The second main advantage of a template is that it can save time. Having significant blocks of text already written will clearly save effort. Having text that is partly written and just needs editing will also be quicker than writing from scratch. And having all the styles and numbering set up will also reduce the labour involved.

This will be advantageous to anyone writing a report. Of course, it will take time to create the template in the first place, so the benefits in terms of time saved will need to be balanced against this – for companies or individuals producing reports relatively frequently, the benefit is likely to outweigh the cost.

What are the drawbacks?

There are disadvantages to using a template. These need to be considered and balanced with the benefits if you're deciding whether to use one or not. Templates also need to be created in a way that 'designs out' some of these, where having a template is still seen as the right approach.

The main problem with templates is that, just as consistency is an advantage, there is a tendency for over reliance on the general or standardised wording in the template, and this can mean that the report is not sufficiently specific.

In other chapters of this book I've talked about the importance of ensuring that a report contains relevant information and is **proportionate** in terms of its length and how certain topics are treated. This means that every report needs to be bespoke, and there isn't really a place for generic text.

Reports written using templates are often much longer than they need to be. An author may feel forced into writing something under a heading, which they don't think is particularly relevant, just because it is in the template.

Templates can take away some of the requirement for an author to think for themselves about the best way to communicate the key messages.

The other major drawback of a template is that if it contains a mistake then the error is likely to appear in every report. And for this reason the template will need to be kept under constant review. A better way of phrasing something might be found, in comparison with how it's written in the template. Or there might be a change to a specific piece of legislation or a guidance document, for example.

The costs associated with creating a template therefore need to consider the ongoing cost of reviewing and updating it. If it isn't updated then the template will cease to be effective and is likely to be ignored.

Other options

The other main way that you might gain some of the advantages of a template is that you start with a previous report for a similar site or similar project, and edit it to work for the current one. This will obviously promote consistency in style, content and layout. It will also help save time, as much of the text will have already been written. And it should prompt the author to include those key pieces of information – unless they forgot them last time as well, or there are fundamental differences compared with the previous report meaning that different information is required. The other advantage is that, assuming you use a very recently written report, you can minimise the chances of including errors stemming from out-of-date references. Of course, if you base your report on one written a little while ago you will be as vulnerable to this as someone writing using a template that hasn't been updated over a similar time frame.

For a very small company or organisation, or someone who is self-employed, this might be a better approach than creating a template.

There are some problems with this approach though. The main one is that it's relatively easy to leave something in the report that was relevant in the previous one, but not in the current one. You therefore aren't immune to including irrelevant information. And what's potentially worse, and likely to be quite embarrassing, is the possibility of leaving in text that is clearly related to a completely different site – this will make it obvious that you've changed a few details and trotted out a previously written report, as well as not having had sufficient attention to detail when it came to proofreading. And, of course, dependent on the details left in, you may have revealed confidential or sensitive information, which could have far more serious consequences. It's a really easy mistake to make – particularly if you have project-specific text in the headers or footers on some or all pages. I've seen this happen many times.

Solutions to the pitfalls of using a template

There is no perfect approach when it comes to deciding on whether to use a template or not – it will come down to personal (or company) preference. However, if you're going to use a template then there are a number of ways to avoid, or at least minimise the likelihood of, falling into the pitfalls. Some of these relate to how you go about using a template (as a report author) and the others relate to how you set up (and revise) your template.

How to set up a template

As I highlighted in the first chapter of this book, different reports have different purposes and different target audiences. This means that they need different structures, different content, and in many cases a different style of writing. For this reason I think it's a mistake to have one report template that you apply to all reports. A company I worked for in the past did this, and it led to a significant number of the problems with the reports that I ended up having to review as a senior member of staff (and, almost certainly, many of the errors I made in reports that I wrote as a junior member of staff).

The first thing you need to decide, when setting up a template, is the type of report you are intending to apply it to. I expect it's possible to create a template that works for more than one type of report. There will be some text that fits in multiple different report types. And if you create more than one template you will have more than one to keep updated. However, the benefits of having a specific template for a specific type of report outweigh any disadvantages, in my view.

Two of the report types most commonly written by ecological consultants are Ecological Impact Assessment (EcIA) Reports and Preliminary Ecological Appraisal (PEA) Reports. Both types will be written often enough to justify the costs of setting up a template. I'm going to discuss these types of report in more detail in Chapter 14. If you're going to use report templates for these reports I would strongly recommend you set up a separate template for each.

You may also find that there are other report types that you often write, and that warrant their own template.

The next thing to do is to come up with a standardised set of section headings and sub-headings for each report type (for which you want to create a template). This should be based on examples of previous reports you've written and the suggested headings and sub-headings set out in Chapter 5 and the relevant appendices.

Once you've selected your headings and sub-headings, you can start to identify the sections that are largely the same each time a given report type is written, with only a few minor edits needed. This might apply to the *Title Page*, the *Introduction* and the *Methods* sections. Text can then be written for each section, with any areas where bespoke text is needed highlighted. This should be done in such a way as to prompt the author to insert the relevant text.

In many cases it will be difficult to provide standard text for the methods section, as each report might relate to different surveys. However, it is possible to write a *Methods* section that covers the vast majority of surveys likely to be undertaken. The author will need to edit those that are relevant in each case and delete those that are not.

The same is true of the *References* section, assuming that you reference documents in that manner, with a list of references at the end of the report. You can have a list including the most commonly used reference sources, with the author deleting those that haven't been referred to in any given case.

Many of the other sections of a report are likely to be more difficult to provide standard text for in the same way. These sections will necessarily be more bespoke, in most cases, and to provide too much 'standard' text in these sections could mislead the author. There's a balance to be struck. One option is to provide instructions for the author on what to include under each heading or sub-heading, and provide some example text to illustrate this. There could also be a series of paragraphs included that are relevant to many reports. For instance, any PEA Report or EcIA Report is likely to include some text relating to the legal protection afforded to nesting birds, the implications that this might have for vegetation clearance, topsoil stripping (for ground-nesting species) or building demolition, and the appropriate approach to minimise the risks.

Clear instructions for the author throughout the template will be incredibly useful. These can be provided as embedded comments, which the author can delete from their report once they have written the text referred to by that particular set of instructions.

In setting up your template, you can also set up the font type and size, line spacing, spacing between paragraphs, and the numbering system for headings/sub-headings and paragraphs. In your word processing software this would be done by creating a named 'style', which can then be easily applied to every report written. You can obviously also set up any headers and footers, the page numbering system, and the automatic contents page.

How to use a template

There are a number of key things for a report author to remember when using a template.

The first rule is, don't overwrite the template – save it as a different document before you start editing.

Secondly, where there's standard text, read it to check it is relevant and correct for the report you're writing. Don't just assume that this is the case.

Thirdly, if you want to change the way something has been written in the template, have a good reason for doing so. Changing text written in the template must be allowed. The standard example in the template cannot possibly always be right. However, don't just change it for the sake of it. The text may well have been written in a certain way for a specific reason. But, if it doesn't work in the scenario you're dealing with, then you have a good reason to change it.

In my view, this could apply to whole sections of a template. As I mentioned when I discussed the drawbacks of templates, the template will encourage the inclusion of irrelevant text, headings and sub-headings. That is, essentially, the trade-off when it comes to templates – as the alternative approach is more likely to omit something. Report authors should therefore be encouraged to delete anything that is clearly irrelevant.

As a very simple example, I read an ecological report recently that had only four paragraphs in the *Methods* section of the report. All four paragraphs appeared under a single sub-heading and a subsequent sub-sub-heading; there were no other sub-headings or sub-sub-headings. So, why have either the sub-heading or the sub-sub-heading? This is almost certainly due to there being standard text provided under other sub-headings/sub-sub-headings in the template report that were deleted because they weren't relevant. This should have been picked up by the report author.

In the same report there was a description of the legislation relating to reptiles. This was necessarily complicated as some UK reptile species are protected under a different piece of legislation from others, and therefore the legal protection has to be described differently for two separate groups of species. Or at least it would have been necessary had species from both groups been relevant to that specific report. In that particular case they weren't – those species protected under one of the pieces of legislation have a restricted distribution and don't occur anywhere near the site that was the subject of the report. Again, standard text was included, and the report author should have edited this to remove the irrelevant information. They might still have needed to explain that there was a separate piece of legislation relevant to other reptiles species, but could have explained that it wasn't relevant in this case, and would then not have gone on to describe it in detail.

There are also a number of things to remember for those 'managing' the template.

The first rule is, save a secret copy somewhere, or make the template 'read only' so that it can't be overwritten and only a copy can be edited. Otherwise, sooner or later, someone will break the first rule of using a template that applies to authors, and will overwrite it. Avoiding this situation, or having a second copy ready for such circumstances will save a lot of time (and probably save quite a lot of shouting as well).

Secondly, you have to keep the template under regular review. Be critical, or ask those regularly writing reports using the template to provide feedback. What could be written better? What would be additional helpful standard text? What has changed that needs updating?

You'll still need to have a specific review after a set period of time, as not all required updates will necessarily have been noticed. And just as you ought to give each version of a report you produce a specific version number, I'd suggest doing the same for your templates. Ideally you should also keep a note of what, specifically, was updated in the template. The template version number could appear in the unique reference number for your reports. This can be helpful when you have to update a previous report for a project – you will then know if it was written using the most up-to-date version of the template and, if not, what has been changed since that report was produced.

Chapter 13

Writing an effective Summary

Having spent a significant amount of time carefully crafting a report, agonising over the specific wording of certain sections, meticulously cross-referencing, checking facts and corroborating opinions, you are entitled to expect that all members of your target audience will read it thoroughly. They will read it from start to finish, noting the correct punctuation, appreciating the in-depth treatment of key issues, whilst admiring the transparent and honest approach to assessment. In short, they will savour every word. Won't they? Well, actually no, they won't. It's a disappointing fact that the majority of the people that receive a copy of your report simply won't have the need, the time or the required attention span to read all of it.

This doesn't mean that you can get away with writing any old rubbish. You can guarantee that your report will get a high level of scrutiny if you do that. Nor should you submit a report that you know has errors in it, on the assumption that nobody will take the trouble to find them. Someone will!

What you should do, though, is write your report in the knowledge that the reader might only be looking for some key pieces of information, taking steps to help them find what they're looking for, without having to wade through lots of detail. There are a number of ways that you might do this, such as by moving lots of technical details to an appendix, or by providing regular summary tables that set out the key messages at the end of certain sections. But the single most important step to take is to provide a *Summary* at the front of the report. This needs to be written in such a way that the reader can understand what the key messages are without reading any further. Of course they may need to read further to allow them to determine what the implications of those key messages are, but the *Summary* should trigger them to do so.

Summaries are therefore very important. They are at least as important as the other sections of the report. They may be the only part of a report that the majority of readers bother to look at. And yet they are often not well written, probably because they are often written last, in a bit of a hurry, when we're thoroughly sick of the sight of the report and just want it to be finished. This is a problem. The *Summary* needs to be effective, and to ensure this means that the author needs to take some time and trouble over it. Often this will seem like a disproportionately large amount of time (and trouble) in comparison with other sections, but that's just the way it is.

So what do I mean by an 'effective' *Summary*? Well, if all members of the target audience can read the *Summary* and make a **sound** decision on whether or not to read the remainder of the report then it's likely to be an effective *Summary*. If the *Summary* doesn't clearly highlight issues of relevance to the target audience, and their lack of awareness of these issues causes them to wrongly decide not to read the rest of the report (or at least certain sections), then it isn't an effective *Summary*.

The *Summary* section is inherently difficult to write. You have to decide what to leave out and what must be included. You have to do this in a way that doesn't hide relevant issues.

So, what are the 'rules' for writing an effective *Summary*?

Here's my first 'rule': it should only be one page long. Well, go on, all right then – a page and a half. More than two pages? Definitely not. Well, there are going to be exceptions of course, but I'd argue that if it's more than two pages long it isn't likely to do its job. Those members of the target audience with very short attention spans won't even read the *Summary*. They might try, but their eyes will glaze over before they get to the end, and their brain will have skipped over some key bit of text. So, **Rule Number 1: keep it brief**.

Rule Number 2: don't write in 'abbreviated' text. What I mean by this is write in full sentences, rather than in shortened sentences, as you might do if you were writing a bullet point. For example:

- *One statutory designated site within search area – SSSI 500m north of site.*
- *Two non-statutory designated sites (CWS) within search area – both 1500m south of site.*
- *Badger setts present.*
- *Great crested newt desk study records.*

The problem with doing this, is that there's a tendency not to explain the context. What is the relevance of these issues and why are they being raised?

Summarising something is different from abbreviating it. The aim is to make a piece of text convey key messages in a succinct way, not simply to make a sentence shorter.

Let's take the designated sites example from my bullet point list above. If one of the key outcomes of a report is to confirm whether a proposed project could affect any designated nature conservation sites or not (which will be the case for PEA Reports and EcIA Reports), then a **clear** statement in relation to this is required in the *Summary*. Some of the key details about the sites are missing from my example text also, such as the names of the sites and, in cases where the assessment predicts possible impacts, the reasons for their designation.

So, for example, you might write:

> *There is one statutory designated nature conservation site within the desk study search area (2km around the site): Blackberry Wood Site of Special Scientific Interest (SSSI) located approximately 500m to the north of the site. There are also two County Wildlife Sites (non-statutory designated nature conservation sites) within the desk study search area: Smith's Heath and Smith's Wood, both located approximately 1500m to the south of the site. None of these sites, nor any other designated nature conservation sites located outside of the search area, are considered likely to be affected by the proposed project.*

Or, if the likelihood of an impact on a designated nature conservation site is clearly very low and this assessment is not particularly controversial, you could express this more succinctly. For example:

> *No statutory or non-statutory designated nature conservation sites are considered likely to be affected by the proposed project.*

Or, if a possible impact on the SSSI has been identified, but not on any other sites, you might write something along the lines of:

> *Blackberry Wood Site of Special Scientific Interest (SSSI) is located approximately 500m to the north of the site. It is designated as it supports ancient woodland. The proposed development may affect the SSSI through increased visitor pressure causing damage to sensitive plant communities. Further consultation with Natural England and the Local Planning Authority in relation to this potential impact is recommended. No other statutory designated nature conservation sites, and no non-statutory designated nature conservation sites, are considered likely to be affected by the proposed project.*

I personally try to avoid using bullet points in the *Summary* except perhaps to list something that doesn't really need further explanation, such as the list of field surveys undertaken perhaps, for example:

> *This assessment is based on a desk study, a habitat survey, and the following further ecological surveys:*
>
> - *Great crested newt eDNA survey;*
> - *Breeding bird survey;*
> - *Bat activity survey; and*
> - *Badger survey.*

Rule Number 3: the text in the *Summary* should work as a stand-alone document. This means that the reader should not be directed to the subsequent sections of the report, its figures or appendices. So for example, you might list the habitat types recorded on site, highlighting those of particular relevance to the assessment, rather than writing *'see Figure 1 for a map of the habitat present on site'*.

Rule Number 4: include a single paragraph summarising the purpose(s) of the report. The reader must be able to read the *Summary* alone and understand the specific purpose (or purposes). In fact that's the first paragraph of the *Summary* pretty much sorted: copy the paragraph from your *Introduction* section that sets out the purpose(s) and paste it into the *Summary*. It may need a minor edit, but in general that's a pretty easy start. Make sure this paragraph includes details of the report's author, as well as details of the client (in other words, the person/organisation that has commissioned the report).

Rule Number 5: include a single paragraph summarising what was done to achieve the purpose(s). This doesn't require much detail, but the reader should get a general understanding of the key elements. So for example, if the work undertaken comprised a desk study, an assessment of the suitability of trees and buildings for use by roosting bats, targeted 'emergence' surveys of specific trees and buildings, a bat activity transect survey undertaken at monthly intervals between May and September inclusive, and the deployment of one automated bat detector for a period of five consecutive nights in each of those months, there isn't going to be space to say much more than that. Perhaps you might add a couple of other key pieces of information, such as what type of automated detector was used, what was the study area for the desk study and the field surveys (did they extend outside of the site), which specific trees and buildings were the subject of 'emergence' surveys. You should also make it **clear** if there were any significant limitations of any of the surveys. And it's worth stating whether or not the surveys accord with the recognised good practice guidance.

What you don't need to include in the *Summary*, in this case, are the specific dates, times and weather conditions of each survey, lots of detail about survey limitations that don't have any major bearing on the outcomes, how many surveyors were on site, and what was their level of competence or did they hold a licence, and how were bat calls recorded, stored and analysed to allow species identification. This information must all appear in the report (in the main text and/or its supporting appendices), but can probably be excluded from the *Summary*.

Clearly some reports will be underpinned by significantly more in terms of surveys than others. An EcIA report, for example, might have involved all of the bat surveys used in the example above, but also a habitat survey, both breeding and wintering bird surveys, reptile surveys, various different amphibian surveys, a badger survey, water vole and otter surveys, and a range of different invertebrate surveys. If a similar level of text is provided in the *Summary* for each of these surveys, as used in the example about bats, there would be several paragraphs about survey methods, and there would be very little hope of fitting the *Summary* onto a single page. In this case I would probably just list the key types of survey and significantly reduce the amount of text for each, such as:

> *The following ecological surveys were undertaken as part of this assessment:*
> - *Habitat survey* [but be explicit about the type of habitat survey]
> - *Breeding and wintering bird surveys*
> - *Reptile surveys*
> - *Amphibian surveys* [you might need to be explicit about which specific surveys]
> - *Badger survey*
> - *Water vole and otter survey*
> - *Surveys of buildings and trees for roosting bats*
> - *Bat activity surveys (transect survey and deployment of an automated detector)*

Where a report is much shorter than my initial example about bats, you could increase the amount of detail in the paragraph in the *Summary* relating to the methods used. However, I'd try not to include additional detail unless you think it is fundamentally important. There's no need to make the *Summary* fill a whole page – half a page is absolutely fine.

Rule Number 6: include any key issues of relevance to the target audience. This is hopefully fairly obvious, but is probably the most difficult part of the *Summary* to write. You have to work out what the key issues are, as far as your target audience is concerned, and boil those down to a handful of paragraphs. Try thinking about the essential pieces of information – what must the reader know if they only read the *Summary*. I normally scroll my way down through the report starting after the *Methods* section and finishing before the *Conclusions*, and write myself a list of points to include. Then I think about the best and most succinct way of writing them into the *Summary*.

Part of the skill here is deciding what is 'key' and what is less important, and giving them **proportionate** treatment. It might well be essential to make the reader aware of the general mitigation measures being proposed in an EcIA report, for example, but some of those measures (the complex ones) will require more text than others (the routine and straightforward ones). Something I often do is to include a brief sentence stating that mitigation measures for X, Y and Z are included in the report, without going into detail about what they are. I think this is acceptable, if the measures are routine and straightforward. And this doesn't break Rule Number 3, as you're not directly referencing the reader to a specific piece of text, you're just making them aware that something is in the report, but you're not going to give more detail on it in the *Summary*. This then allows you more space to provide the detail on issues of greater consequence, bearing in mind the purpose(s) of the report and the conclusions that you are attempting to reach.

And that brings me neatly on to **Rule Number 7: include a single paragraph summarising the key conclusions**. The reader must be able to read the *Summary* alone and understand the conclusions of the report, so this should be the final paragraph of the *Summary*: copy the paragraph(s) from your *Conclusions* section and paste that into the *Summary*. It may also need a minor edit, particularly if you've written a lot in that section and need to condense it into a single paragraph.

Rule Number 8: everything written in the *Summary* must also be included in the main body of the report. This includes any piece of survey data, any bit of background information taken from a reference source, and any opinion. When you start to write a *Summary* you might find yourself thinking differently about how to express a particular opinion, and that might lead you to include new pieces of information that weren't included in the main report. This shouldn't happen – the *Summary* is simply a shortened version of what's in the report. Imagine if the abridged version of a novel included a whole series of new characters that didn't appear in the original!

Rule Number 9: make it appeal to the lay reader. You should write the *Summary* with an audience of non-ecologists in mind. So you should avoid some technical information or terms. I wouldn't include the Latin names of any species I refer to in the *Summary*, for example, as this might mean that a lay reader has to stop and read a sentence several times, with little (if any) additional benefit. This might mean excluding some details that an ecologist would have been perfectly comfortable reading, but remember that most readers of the *Summary*, and certainly most of those people who don't read past the *Summary*, will be non-ecologists.

You do have to be careful not to exclude technical information that is of particular importance, as this might been seen as misleading the reader. For example, if a Site of Special Scientific Interest is of relevance to an EcIA report, and is therefore discussed in the *Summary*, you couldn't avoid using the term 'Site of Special Scientific Interest'. Whilst it might be a term that some non-ecologists will be less familiar with, it is fundamentally important that the reader is informed of the specific designation of that site, as this term is used in planning policy for example.

And finally (appropriately) here's my **Rule Number 10: write the *Summary* last**. You should have completed the remainder of the report before you start writing the *Summary*. It's tempting to start filling in certain sections/ paragraphs before you've completed some elements of the main report. I'd advise against doing this. You need to make sure that the *Summary* is based solely on information provided in the main report, that it gives a balanced treatment of the key issues, and is accurate. Starting to write the *Summary* before you've completed everything else is risky – there's a chance that something will slip through the net.

And just to be really helpful, here's a summary, as a reminder of what I've covered in this chapter:

Rules for writing an effective Summary

1. Keep it brief
2. Don't write in 'abbreviated' text
3. The text in the *Summary* should work as a stand-alone document
4. Include a single paragraph summarising the purpose(s) of the report
5. Include a single paragraph summarising what was done to achieve the purpose(s)
6. Include any key issues of relevance to the target audience
7. Include a single paragraph summarising the key conclusions
8. Everything written in the *Summary* must also be included in the main body of the report
9. Make it appeal to the lay reader
10. Write the *Summary* last

Chapter 14

PEA or EcIA Reports – what's the difference?

In the first chapter of this book I highlighted the importance of understanding, and clearly describing, the purpose of the report. Remember Key Characteristic 1? It was being **purposeful** – a report needs to have clear aims and objectives that meet the intended user's expectations and that it delivers against.

I also stressed how critical it is that the author of a report knows the answer to these two questions:

1. What is the purpose of the report?

2. Who is the target audience for the report?

Box 3 in Chapter 1 sets out some of the different reports that a professional ecologist is likely to have to write, along with their various purposes and target audiences. In this chapter I'm going to focus on two of those reports in detail – Preliminary Ecological Appraisal Reports (often referred to as PEA Reports or PEARs) and Ecological Impact Assessment Reports (which I'll shorten to EcIA Reports). The reasons that I want to focus on these two types of report in particular are that:

1. They are probably the reports that ecological consultants most frequently write;

2. They are key to the process of designing developments and applying for planning consent, which means that there is (quite rightly) a high level of scrutiny applied to these reports, and the consequences of poor quality reports (see Box 1 in Chapter 1) are particularly acute; and

3. There are some overlaps between the two, mainly because they are both produced in relation to development projects and are produced before planning consent is granted – which means that they are often confused with each other, and this confusion has led directly to a good number of the examples of poor quality reporting that I've seen.

Incidentally, CIEEM has published guidelines on both PEA[1] and EcIA[2] – these guidelines should be referred to as they describe the process of PEA and EcIA, but also provide some guidance on how to report the outcomes of those processes (which is what I'm going to discuss in the following sections).

PEA Reports

Let's start by looking at PEA Reports, in terms of their purposes and target audience. A PEA Report will normally have the following purposes (as set out in Box 3 in Chapter 1):

1. *To identify ecological constraints to a particular development scenario, as well as opportunities for delivering enhancement or biodiversity benefits;*

2. *To identify the mitigation measures and licences likely to be required; and*

3. *To identify any further surveys needed.*

---

1 CIEEM (2017) *Guidelines for Preliminary Ecological Appraisal, 2nd Edition*. Chartered Institute of Ecology and Environmental Management, Winchester.

2 CIEEM (2018) *Guidelines for Ecological Impact Assessment in the UK and Ireland: Terrestrial, Freshwater, Coastal and Marine*. Chartered Institute of Ecology and Environmental Management, Winchester. Version 1.1 updated September 2019.

And the target audience for a PEA Report is normally a developer, as well as members of their team involved in designing that development. Importantly, when a professional ecologist writes a PEA Report for a developer, they essentially become part of that design team. They are advising the developer on how he or she can achieve their objectives. It is also the case, of course, that they must be unbiased and objective (or **impartial** as I've called it in Key Characteristic 8) and have a duty to the natural environment as well as to their client – because anything else would be considered to be unethical and, for members of a professional body like CIEEM, because they have signed up to a code of conduct that requires this. Balancing these, sometimes competing, objectives effectively takes experience and is a key part of being an ecological consultant.

A professional ecologist might see their role in undertaking a PEA as trying to communicate the impacts of a particular development project on ecological features to a developer. However, that's not really the information that a developer wants to have communicated to them (although this information is vitally important and will need to be communicated to others, on a developer's behalf – see EcIA Reports, below). What the developer wants to know is what is the impact of ecology on their development? And when I say 'on their development' what I mean is 'on their ability to deliver a given development project to a specific programme and a specific budget'. Anything that threatens this is likely to have a developer sitting up and taking notice. Anything that doesn't is really just detail that they aren't overly interested in – at least not until they have planning consent.[3]

The PEA Report is one of the means that a professional ecologist will communicate these 'big issues' to a developer. The information in the report will be used to help guide a client's fundamental decisions about a project, such as:

- Is development of the site viable?
- Is a development of a site likely (or unlikely) to be granted planning consent, given the relevant policies and the ecological features potentially affected?

3 I'm not condoning this view, by the way – just using it to highlight the importance of understanding your target audience and their perspectives.

- How should the development be designed to be 'acceptable' in terms of its impacts on ecological features?

The information on mitigation measures and licences likely to be required will be important in ensuring that the developer understands the likely costs and the programming implications. In most cases it will not be possible to be completely explicit about the mitigation measures (and any licences) needed, as this could be affected by design changes and the results of any further surveys – this information needs to be expressed in a way that makes this **clear**, ensuring that worst-case scenarios are covered.

And finally, the identification of further surveys needed should ensure that the developer is aware of what other ecological work needs to be completed before they can apply for planning consent, and what the programme implications are of this further work. It is vital that this is communicated effectively in the report so that a developer can commission further surveys in a sufficiently timely manner.

So, in summary, a PEA Report is written for a developer to read, informing them of the things that they need to be aware of – key design constraints, any fundamentally important 'big issues', anything that could have an impact on their programme or budget. It is therefore likely to contain recommendations, general advice, and possible options to be 'considered' – none of these things should feature in an EcIA Report.

It is also likely that the design of the development has not yet been finalised at the time that a PEA Report is written – that's the point, it's meant to help guide the design (although, admittedly, this doesn't always happen). And it's possible (and likely in most cases) that additional ecological information will be needed, beyond that available at the time of writing the PEA Report, before a planning application can be submitted.

EcIA Reports

EcIA Reports, then, have a completely different set of purposes from a PEA Report as well as a different target audience. An EcIA Report will normally have the following purposes (as set out in Box 3 in Chapter 1):

1. *To present an assessment of the likely (significant) effects of a development proposal on ecological features/biodiversity;*

2. *To allow a determining authority to ascertain whether the proposal accords with relevant planning policy and legislation; and*

3. *To allow a determining authority to write planning conditions or obligations (where necessary) to secure mitigation, compensation and enhancement measures.*

The main target audience is hopefully fairly obvious from the purposes – it is the authority responsible for determining whether the development should go ahead or not (normally the Local Planning Authority). This might include a biodiversity officer (a fellow professional ecologist), a case officer (normally someone with a background in planning), and members of the planning committee (who will come from a variety of backgrounds). In making their decision they are likely to seek the views of the relevant Statutory Nature Conservation Body as well as non-statutory nature conservation consultees (such as the local Wildlife Trust). Consultees such as these are therefore also part of the target audience.

There will be a number of other interested parties in any development project, including local residents. Any information submitted as part of a planning application will be made publicly available. Members of the public therefore also form part of the likely audience – although the report will not be written specifically for them as individuals, the author needs to be aware that at least some members of the public are likely to read it with a view to gaining specific information from it. They should therefore be treated as part of the target audience as well.

When discussing PEA Reports earlier in this chapter, I said that a developer wants to understand the impact of ecology on their development (and this is the role of the PEA Report). I also said that professional ecologists often forget this and instead try to communicate the impacts of a particular development project on ecological features to a developer (which is not the information they want). However, this is exactly what the determining authority need to allow them to make a sound planning decision – an understanding of what the implications are of them granting consent. This is the role of the EcIA Report.

The EcIA Report is therefore not giving advice to a developer. It is written once the developer has designed the development and all further surveys needed have been undertaken. It explains the likely outcomes for biodiversity of a project (if it is granted consent) and identifies the measures that need to be secured as part of any consent (or through a protected species licence, for example) to ensure that the project accords with planning policy and is legally compliant.

The design therefore needs to be fixed to allow an EcIA Report to be written – there may be some elements of detail still to be determined, but these should be accounted for in the report.

And, importantly, all ecological surveys needed to allow the report to achieve the purposes set out above must have been completed, and the information provided in the EcIA Report. Of course, in some cases there may be no additional ecological surveys required beyond what was done for the PEA Report – you don't have to have done further ecological surveys in order to write an EcIA Report.

An EcIA Report should therefore not include recommendations or general advice. There may be scenarios where multiple options can be assessed, but these need to be specific.

What do these differences mean?

The different purposes and target audience of PEA and EcIA Reports mean that as well as having different contents they should have a different structure (or, at least, a partly different structure) and they should also have a different emphasis.

There is likely to be some overlap in terms of content and structure:

✓ Both will contain a *Summary*;

✓ Both will have an *Introduction*, setting out the purpose of the report (although the purpose will be different);

✓ Both may have a section setting out the relevant planning policy and legislation;

- ✓ Both will have a *Methods* section – in both cases this is likely to include a description of a desk study and a basic habitat survey and assessment;
- ✓ Both will include details of designated nature conservation sites within the Zone of Influence of the project;[4]
- ✓ Both will include details of the habitats present on site and in the surrounding area; and
- ✓ Both will include references to relevant good practice guidelines.

However, there will be some major differences as well:

- ➢ An EcIA Report will need to include a section on *Assessment methods;*
- ➢ An EcIA Report will include detailed methods and results of further ecological surveys, which would have been recommended in a PEA Report (assuming such surveys are required);
- ➢ An EcIA Report will normally focus on ecological features of relevance to the assessment of effects, 'scoping out' features unlikely to be affected or not of relevance to the determining authority, to allow the report to focus on the main issues;
- ➢ An EcIA Report will normally describe the baseline conditions for, and assign a scale of geographic importance to, the ecological features likely to be affected – a PEA Report may also include this, but may struggle to do this for all features prior to the completion of those further surveys;
- ➢ An EcIA Report will be based on a specific fixed scheme design, rather than making recommendations for how the design could be modified, and should also include a *Scheme description* section; and
- ➢ An EcIA Report will contain <u>specific</u> mitigation <u>proposals/commitments</u> that are intended to deal with identified impacts, whereas a PEA Report will normally contain <u>generic</u> mitigation <u>recommendations</u>, as the design may not have been fixed at the time of writing.

---

4 Defined by CIEEM in their Ecological Impact Assessment Guidelines as 'the area(s) over which ecological features may be affected by the biophysical changes caused by the proposed project and associated activities' (CIEEM 2018; see footnote 2 earlier in this chapter for the full reference).

As you can probably see from the differences I've highlighted, the emphasis of a PEA Report is about advising and recommending, with generalities rather than specifics. The emphasis of an EcIA Report is about committing to a specific approach, in terms of design and mitigation.

These fundamental differences are the reason that, in most cases, a PEA Report should not be submitted with a planning application (unless an EcIA Report is also submitted). It won't normally provide the level of certainty that a determining authority requires in order to make an informed decision on a planning application.

A PEA Report should therefore be treated as an internal document between a developer and their ecological consultant, or could be used as a means of agreeing the scope of work required for an EcIA with the determining authority.

When can you submit a PEA Report instead of an EcIA Report with a planning application?

There are circumstances when a PEA Report can be submitted with a planning application instead of an EcIA Report. These are when:

1. *No further ecological surveys are needed*

   For very straightforward sites, dependent on the habitats present, it may be possible to base an impact assessment on a desk study and habitat assessment only.

   AND

2. *There is sufficient detail available in relation to scheme design to allow a robust and reliable conclusion on the likely effects to be drawn*

   This will be the case where either a fixed design is available at the time that the PEA Report is produced, and no changes are needed to avoid/minimise significant effects, or where the site is so lacking in ecological interest that it doesn't matter how the scheme is designed – provided that it is located within the 'site' the effects would not be significant.

   AND

3. *There is sufficient information within the report about ecological mitigation (and compensation) as well as enhancement measures to allow the certainty needed by the determining authority*

   In other words, the determining authority can secure any necessary mitigation, compensation and enhancement measures on the basis of the PEA Report.

Where this set of circumstances exist, a PEA Report can be submitted instead of an EcIA Report, and should be sufficient to allow the determining authority to make a decision on the application. If this is the intended use of the PEA Report then this should be made clear in the *Introduction*, and the emphasis of the report should be closer to that of an EcIA Report than that of a PEA Report (as there will be nothing to recommend and it will be necessary to be more specific about mitigation, compensation and enhancement measures). Alternatively, the report could be produced following the structure and content of an EcIA Report.

In 2019 CIEEM and the Association of Local Government Ecologists (ALGE) produced a Checklist for EcIA Reports,[5] to allow those writing such reports to check that they contained the necessary information, and to allow those reviewing them to do likewise. This can be downloaded from CIEEM's website. One way of checking that a PEA Report is suitable to accompany a planning application, effectively doing the job of an EcIA Report, is to review it against the criteria in the Checklist – if it contains everything that an EcIA Report should (and the three circumstances described above all apply), then the PEA Report should be sufficient. I'll discuss this Checklist again in Chapter 16, when I look at proofreading, technical review and Quality Assurance.

---

5 CIEEM/ALGE (2019) *Ecological Impact Assessment Checklist*. Chartered Institute of Ecology and Environmental Management, Winchester and the Association of Local Government Ecologists. https://cieem.net/resource/ecological-impact-assessment-ecia-checklist/

Chapter 15

Writing Environmental Statement chapters

Background

An Environmental Statement is the document produced setting out the results of an Environmental Impact Assessment (EIA) under the EIA Regulations. At least this is what that document is called in England, Scotland and Wales. The equivalent document is called an Environmental Impact Statement in Ireland, and will be called something different in other parts of Europe. The document comprises a series of chapters. Each environmental discipline will have its own chapter – these might include, for example, 'Landscape and visual', 'Hydrology', 'Archaeology' or 'Cultural Heritage', 'Noise', or 'Air quality'. The chapter dealing with ecological issues should now be called 'Biodiversity', as this is the topic identified in the latest version of the EIA Regulations, although it used to be termed 'Flora and fauna'. You will regularly also see chapters termed 'Ecology'. And in some cases you might see more than one chapter dealing with ecological issues – a proposed wind farm might have a separate chapter dealing with impacts on birds, termed 'Ornithology' or 'Avian ecology', separate from other ecological impacts.

What is different about writing a chapter of an Environmental Statement, as opposed to any other ecology report? Well, the short answer is 'not that much'. The same general rules apply. Environmental Statement chapters are, effectively, no different from an Ecological Impact Assessment Report. They have the same purposes and same target audience. They need to be **robust**. They need to contain all relevant data to allow proper scrutiny. They need to be **clear** and easy to read, even for someone without specialist ecological knowledge. They have to be based on sound professional judgement, with supported opinions, and written by somebody who is **competent**. All of these things apply to an EcIA Report for a non-EIA project as well as to an Environmental Statement chapter.

The main differences between an EcIA Report and an Environmental Statement chapter relate to the simple fact that an EcIA Report is a stand-alone document, whereas an Environmental Statement chapter is obviously part of a larger document. The structure of an EcIA Report is therefore entirely up to the author (or the company or organisation that the author works for). The structure of each Environmental Statement chapter, on the other hand, is normally dictated to the author, to ensure that it is consistent with the other chapters in the document. This has a number of implications, which we'll look at in turn.

Scheme description

Firstly, an Environmental Statement will have a 'Scheme Description' chapter. This means that there's no need to include a section describing the proposed project in the chapter dealing with ecological impacts (although there would need to be such a section in a stand-alone EcIA Report). The Scheme Description chapter is normally written by someone involved in the overall co-ordination of the document – an EIA Co-ordinator in many cases. Whilst this might seem like that's one less thing for the ecologist to write, this does mean that the author of the Ecology or Biodiversity chapter needs to have the Scheme Description chapter in front of them when they write their chapter, along with the associated drawings or figures. This is important, as otherwise they might be assessing ecological impacts in relation to incorrect specific details of the project. The key, then, is to make sure you get a draft version of the Scheme Description chapter before you start writing your own chapter.

Assessment methods

Secondly, an Environmental Statement will also have an 'Assessment Methods' chapter. This chapter will describe the approach being taken to assessing impacts and determining significance of effects. A stand-alone EcIA Report will normally have a section describing this in the *Methods* section. An Ecology or Biodiversity chapter of an Environmental Statement will need to cross-refer to the Assessment Methods chapter. So this is another chapter that you'll need to have a draft of when writing your own chapter.

It's also worth highlighting here that, quite often, the assessment methods being used for ecology (or biodiversity) will differ from those being used for the other environmental disciplines. This is because the Chartered Institute of Ecology and Environmental Management (CIEEM) has published guidance on Ecological Impact Assessment,[1] which differs from approaches that are widely used in EIAs for other topics. So it is possible that the Assessment Methods chapter may describe an approach that is being used by the Landscape assessment, or Archaeology, or Noise, for example, but is not being followed in the Ecology or Biodiversity chapter. If this is the case it will need to be agreed with the client or EIA Co-ordinator first. This isn't the place to go into the advantages of one approach over the other. However, I do need to highlight the importance of clarity on the assessment methods being used. So, if you're writing the Ecology or Biodiversity chapter of an Environmental Statement, you will be in one of two basic camps:

1. You are following the approach to assessment exactly as set out in the Assessment Methods chapter. In this case you will not need your own section describing the assessment methods being used as you can simply cross-refer to the chapter that does this already. You will, of course, need a draft of this chapter in front of you when writing your own chapter.

2. You are largely following the approach to assessment set out in the Assessment Methods chapter but modifying it in some way, or are following a completely different approach (such as that described in the CIEEM Guidelines on Ecological Impact Assessment). In either case you will need to include your own description of assessment methods in the Ecology or Biodiversity chapter and explain very clearly that they differ from those described in the Assessment Methods chapter, along with an explanation of how they differ and why. The Assessment Methods chapter should also be modified to ensure that this is **clear**, with an appropriate cross-reference to the text in the Ecology or Biodiversity chapter.

---

1 CIEEM (2018) *Guidelines for Ecological Impact Assessment in the UK and Ireland: Terrestrial, Freshwater, Coastal and Marine*. Chartered Institute of Ecology and Environmental Management, Winchester. Version 1.1 updated September 2019.

Structure

Thirdly, the structure of an Ecology or Biodiversity chapter of an Environmental Statement is often different from a stand-alone EcIA Report, in a couple of ways in particular:

1. There will normally be separate sections describing the potential ecological impacts of a proposed development, describing the mitigation measures, and setting out the significance of residual effects. Each of these sections will then need to be subdivided for each ecological feature being considered (each different designated site, habitat or species population of relevance to the assessment).

2. The potential impacts, and sometimes also the mitigation, are usually split into different 'phases', such as 'construction phase' or 'operational phase' or any one of a number of other phases, dependent on the type of project.

A stand-alone EcIA Report can be set out in the same way, but the structure that I've recommended in Appendix C has the potential impacts, mitigation and residual effects for each ecological feature all in one section (under the sub-heading of that particular feature). This is because it tends to flow more logically if it is set out in this way, and there's less chance of repetition. The Environmental Statement chapter approach has an advantage in that it describes all the mitigation measures in one part of the report, which is helpful for a determining authority when it comes to writing planning conditions. An EcIA Report using the layout shown in Box 6 can achieve the same outcome by including a summary table at the end of the section, listing all mitigation measures.

Cross-referencing

Consistency between chapters of an Environmental Statement is important, in terms of design information being used, and in terms of some of the mitigation measures. If some of the ecology mitigation is being delivered through new landscape planting, for example, this will need to be discussed in both the landscape and ecological assessments. It's obviously important that an Ecological Impact Assessment, whether part of a stand-alone EcIA Report or as part of an Environmental Statement chapter, is written with

Box 6: Typical layout of assessment section of an Environmental Statement chapter versus a stand-alone EcIA Report

Environmental Statement chapter	EcIA Report
Potential construction phase impacts • Feature A • Feature B • Feature C • Etc.	*Feature A* • Potential impacts (can be subdivided into construction phase, operational phase, etc.) • Mitigation measures • Significance of residual effect
Potential operational phase impacts • Feature A • Feature B • Feature C • Etc.	*Feature B* • Potential impacts • Mitigation measures • Significance of residual effect
Mitigation measures • Feature A • Feature B • Feature C • Etc.	*Feature C* • Potential impacts • Mitigation measures • Significance of residual effect Etc.
Significance of residual effects • Feature A • Feature B • Feature C • Etc.	

Note: other phases beyond 'construction' and 'operational' phases may also need to be included.

reference to other related topics in the assessment. In a stand-alone EcIA Report this consistency will need to be achieved by obtaining, reviewing and cross-referencing the reports produced by other specialists. In an Environmental Statement the cross-reference will need to be between chapters in the same document. This can be difficult to achieve, particularly if all the chapters are being written at the same time.

Consultation

The Environmental Impact Assessment process will normally involve a reasonable level of consultation with the determining authority and possibly also with nature conservation consultees, through production of a scoping report for example. This will mean that there is often more to report in relation to the methods and outcomes of consultation. Any consultation responses received, and how the comments have been dealt with, can usefully be set out in an Ecology or Biodiversity chapter of an Environmental Statement. Alternatively, they may be dealt with in one location within the Environmental Statement for all environmental disciplines combined.

Summaries

The summary in an Environmental Statement is normally provided as a single document, which provides a summary for each environmental discipline. This means that the summary text for ecology (or biodiversity) needs to work as a stand-alone document. It should not be written in abbreviated or bullet point form. This is, in any case, good practice for an EcIA Report or, indeed, any other ecological report (see Chapter 13).

Chapter 16

Proofreading, technical review and Quality Assurance

So, you finally finish your report. You've slaved away over it for days or weeks (or perhaps longer?). Frankly, you're sick of the sight of it and can't wait to send it to the client, because at least then the misery will be over. And then someone has the audacity to tell you that it needs proofreading or checking or something else! Well, let's hope so anyway. Sending out a report without a thorough check is a massive mistake.

Many companies or organisations will have their own Quality Assurance processes. Smaller companies or sole traders may not. Either way, there are some key steps that I would recommend when checking a report, which should fit into any existing processes (not replace them). You may find that you're doing these things already, which is great. You may find that you're doing additional things as well, to be in accordance with your Quality Assurance processes, which is even better. Even if you are doing these steps already I've included some tips on how to do them, so it's worth reading on!

Step 1: The author proofreads the report

I know that this will be incredibly annoying because, having just finished the report, this is the last thing you want to be doing. However, it's really important for two reasons:

1. There will be mistakes in the report and some of these may not be that easy for someone else to spot.

   Bear in mind the remaining steps and who is going to undertake them. For example, will they pick up the fact that Woodland A is to the east of the site rather than to the west, as stated in the report? Well, they might, but then again they might not. As the report author you're likely to notice some errors that someone else won't.

2. Not to do so is, in my view, lazy.

   As the author of a report, with your name on the front cover, you need to take some pride in your work. So, make it as good as you can. I know that someone more senior might start rewriting bits of it, but that doesn't mean you can leave it to them to correct all of the mistakes. Some effort on your part to pick up as many typos and grammatical errors as possible is, as far as I'm concerned, the least I would expect. I can tell you from personal, bitter experience that there are few things more annoying than reading a report that is littered with basic errors. It will take longer for those who have to perform Steps 2 and 3 to complete their part in this. And if there are too many errors still in it when they come to read it they might not be able to complete those steps effectively, with the result that it comes back to you for further correction.

So, it is essential that the author proofreads their own report before they pass it on for Step 2. And when I say proofread that isn't a quick skim through the document. You have to read every single word. That's cover to cover, including the key and the labels on the figures, the data tables in the appendices and the text in the footers. Yes, it may be a little bit boring, but it has to be done. And sooner or later you'll find yourself reading your own report and feeling a sense of pride that, actually, you've produced a good piece of work. Trust me – it's worth the effort.

During the proofreading exercise you're obviously looking for any errors, but you're particularly focusing on the following:

- Typos;
- Incorrect punctuation;
- Grammatical errors;
- Inconsistency in numbering, heading or sub-heading styles, order of headings/sub-headings, format, font size and type, bullet point type, and line or paragraph spacing;
- Missing figures, tables, photos or appendices – where one is referred to, check it to make sure it includes what it is supposed to and that it is numbered correctly;
- Incorrect cross-references – follow each cross-reference to make sure it is correct; and
- Inconsistency in terms used (I've provided some examples at the end of the chapter)

The problem with proofreading your own report, though, is that your brain will skip over the text, to a certain extent, and you may find yourself reading what you thought you'd written rather than what you actually wrote. This is inevitable, and means that the author is unlikely to pick up all of the mistakes (after all, if they did routinely pick up every single mistake there would be no need for Step 2). There are a couple of things I do (or don't do) to try to combat this problem:

1. Don't proofread as soon as you've finished writing

   Take a break, leave it until after lunch, come back to it the following morning, do some other unrelated tasks first – whatever it takes to clear your mind from the writing process.

2. Don't proofread in a hurry

   We're always up against a deadline, or have to head out to do a survey or attend a meeting, and so there's plenty of temptation to rush this crucial stage. But if you do there will be little point in doing it at all. It will take as long as it takes, so choose a period in your day to do this when you don't have to be somewhere at a certain time.

3. Proofread away from the place where you wrote the report (i.e. not at your desk)

   This isn't essential but I find it really helps. It's a change of scene that enables you to focus your mind on the task of proofreading and distinguish it from the previous task of writing. It also helps with the next one!

4. Proofread away from distractions

   Switch off your phone, don't look at your emails, move away from your desk (see above) and find a location where you won't be pestered by colleagues.

5. Proofread a hard copy, not on a screen

   Personally, I think it really helps to proofread a hard copy. This is for several reasons. Firstly, you can more easily move away from your desk and other distractions. Secondly, there will be some errors that only show up on a hard copy and are less obvious on the screen. Thirdly, your mind will be more likely to skip ahead and not read thoroughly if you read on screen. And finally, if you proofread on screen you will end up correcting as you go; this may be fine in some cases, but the problem is that you will be more likely to lose the thread of what you were reading, and you might miss other errors as a result.

Mark up the errors on the hard copy in a coloured pen so that they stand out from the type. Then make the necessary corrections in the electronic version, produce a nice, shiny, new, clean and corrected version, and hand it over for Step 2.

Step 2: Another person, not involved in writing the report, proofreads it

It will be difficult for the author of a report to spot all of their own mistakes. A fresh set of eyes can be incredibly useful. Is it essential? Well, perhaps not for short, simple reports or for reports where the author has a lot of experience. But for longer, more complex reports, or for reports written by someone who doesn't have considerable experience of report writing, I'd argue that it is essential.

Who should do the proofreading? Well, it needs to be someone with a good grasp of grammar, spelling and punctuation. It should be someone who produces good written work themselves, is diligent and has an eye for detail. It doesn't have to be an ecologist, although it can help if it is. And if a non-ecologist does this then it can be beneficial if they are proofreading ecological reports regularly, so that they become accustomed to some of the terms and approaches used.

It also doesn't need to be done by someone with a particular level of seniority – it could be done by a recent graduate, provided that they have the necessary skills.

What are they looking for? Well, they're focusing on spotting exactly the same mistakes as the author.

And the tips for how they should proofread are basically the same as for the author as well, discounting those that are irrelevant as they won't have just finished writing the report, leaving us with:

- Don't proofread in a hurry
- Proofread away from distractions
- Proofread a hard copy, not on a screen

Step 3: A senior member of staff undertakes a technical review

The technical review needs to be completed after the proofreading has been done, so that the person undertaking it is not distracted by basic errors. You might work for a company or organisation that has a different term for this, but as far as I'm concerned the technical review is a check to make sure that the report is 'fit for purpose'.

This part of the process does need to be completed by an experienced member of staff. I recall hearing about a scenario a few years ago, when a client was not happy with the ecological report they had received from their consultant. On closer examination of the report it became clear that the ecologist who had completed the 'technical review' – essentially signing the report off as 'fit for purpose' – had very little experience and was much more junior than the person who wrote the report. This will have undoubtedly led to some of the problems with the report, which a more experienced ecologist would have been likely to spot. This is not necessarily the fault of the individual who had done the technical review (they shouldn't have allowed themselves to be put in that position but that can be difficult if undue pressure is applied and, let's face it, you don't know what you don't know). The fault here is really with the company processes, which should stop this happening, and possibly also with the report author, who should have known better.

The name of the person who undertook the technical review should, in my view, appear in the report, along with an explanation of their competence. If they are signing off the report as fit for purpose then the reader is entitled to know who they are and what makes them **competent**. The same is the case for authors, but I wouldn't apply the same to the individual who did the 'Step 2' proofread – I think it's down to the company or organisation producing the report alone to make sure that they are sufficiently **competent**.

So, what should the person doing the technical review focus on? Well, this is going to differ for different types of report, making it difficult to come up with a standard list that works in all cases. However, it is possible to produce a standard list for specific report types. In 2019 CIEEM published the 'EcIA Checklist',[1] which effectively provides a starting point for a standard list

---

1 CIEEM/ALGE (2019) *Ecological Impact Assessment Checklist*. Chartered Institute of Ecology and Environmental Management, Winchester and the Association of Local Government Ecologists.

for undertaking technical reviews (or critical reviews – see Chapter 17) of EcIA Reports. It provides a series of criteria to be used in judging whether an EcIA Report is fit for purpose.

It's worth either using the EcIA Checklist for the technical review of any EcIA Report, or reviewing your own existing list and incorporating any additional criteria from the EcIA Checklist into it.

It could also be adapted to provide a parallel checklist for PEA Reports, for example.

Along with the specific criteria it identifies, the other key thing about the EcIA Checklist is that it asks the person completing it to identify the paragraph or section of the report that shows compliance with each of the criteria. This means that it isn't a simple 'tick box' exercise. You can't just go down the list ticking everything without demonstrating compliance. To do that would be a pointless exercise. So, if you're going to develop a specific list of criteria for undertaking technical reviews, make sure that the person doing the check can't get away with simply ticking the boxes.

There are a number of general questions that the technical reviewer needs to be considering when undertaking this task (and I'm sure there are others besides these as well):

1. Is the report appropriately titled, with a date, reference number and version number provided on the *Title Page*?

2. Does the report meet the client's brief?

3. Is the purpose of the report clearly stated?

4. Has the purpose of the report been met?

5. Is the report appropriately structured, given its purpose and intended target audience?

6. Are the individuals involved in producing the report, or undertaking surveys that the report is based upon, appropriately 'qualified', and are their names and competencies clearly stated?

7. Are the methods and limitations clearly described?

8. Do the methods follow good practice guidelines, or clearly state and justify any departures from such guidelines?

9. Are the desk studies and field surveys up to date?

10. Have all relevant desk study and field survey results been included to a sufficient level of detail?

11. Are the figures clear and easy to interpret?

12. Are any major issues flagged appropriately?

13. Have any possible issues relating to legal compliance been adequately addressed (this might relate to statutory designated nature conservation sites, protected species, or invasive non-native species)?

14. Are any recommendations or proposals appropriate and justified?

15. Have any seasonal constraints been clearly explained (in relation to further surveys or mitigation measures, for example)?

16. Are the *Conclusions* clearly stated and do they relate to the purpose of the report?

17. Are the *Conclusions* substantiated by the text?

18. Have appropriate reference sources been used, and are they acknowledged appropriately?

19. Does the *Summary* provide a clear message of the key points of interest to the target audience?

20. Has any information that should remain confidential been treated appropriately, or removed?

Combining the steps

Can the report author undertake the technical review? Well, ideally they shouldn't, because it will be difficult for them to satisfy themselves of the answers to these questions in an unbiased way – they have a vested interest as they wrote the report.

I'm not going to tell you that the author can't do the technical review, as that will be practically difficult to achieve in some cases, such as for sole traders or small companies where there are relatively few senior staff members. I'm also not going to tell you that the author can do the technical review, firstly because it's clearly not ideal and secondly because it might not be allowed by your company or organisation's Quality Assurance processes.

It's up to individual companies or organisations to make their own rules on this. The technical review does need to be a thorough exercise, and undertaken by someone with the right level of expertise. So, where a company or organisation does allow report authors to also do the technical review I would advise that this is limited to the simpler reports, or those with restricted purposes, where errors are less likely to be made and will have fewer consequences. More complicated reports, or reports with major consequences if they are incorrect, should probably be treated differently.

Can Steps 2 and 3 be done by the same person? Well, in my view, yes, they can (although check your company/organisation's Quality Assurance processes allow this first). For reports that are well written in the first place, and have had a good proofread by the author, an experienced technical reviewer may feel confident in performing Steps 2 and 3 at the same time. Where the report is littered with errors, this will be more difficult. And someone used to doing technical reviews will find it difficult to do a Step 2 proofread without also undertaking elements of the technical review. So whilst it might be difficult to disentangle the two steps, they'll probably have to read the report twice.

Other checks

There may well be other checks included in any given Quality Assurance process that I haven't mentioned here and these will obviously need to be completed.

One possible other check that is sometimes appropriate is for the staff involved in undertaking the surveys to read the report, or certain sections of it, to check that their findings have not been misrepresented. This obviously doesn't apply to reports written by the person who did the survey, but may be relevant in some cases, and will be particularly relevant for companies that use different people to do surveys than to write reports (which is not something that I would recommend).

A way of achieving this efficiently might be to ask the surveyor to do the Step 2 proofread, assuming that they are sufficiently diligent and produce good written work themselves.

Documenting the checks

It's important that the various steps described above are documented. This will be a key part of any Quality Assurance process. Again, if you have such a process then I'm not going to tell you to do it differently. If you don't then I'd strongly recommend that you record the name of the person that undertook each of the steps in the checking process and the date when they undertook their step. Many companies' processes will also require a signature from the relevant person, and will include this on the Title Page or inside cover of a report.

Whatever approach is taken to this it's important that a record is kept.

It is also worth keeping a record of the original versions of the report, before amendment. This might be the printed hard copy, or could be a protected electronic version. Version control will be important for this. And if you're going to keep electronic versions of drafts before they go to the client, I'd think carefully about how these are labelled/numbered to make sure that:

1. A draft version isn't issued accidentally
2. The client doesn't receive a report identified as version 3, for example, when they've never seen a previous version

Classic mistakes to look for

There are some errors that are regularly made in ecological reports, particularly when it comes to consistency of terms. For example:

1. Style rules for common names and Latin names of species (see Chapter 3 for more detail on this):
    - Latin name in brackets, or not?
    - In italics and at first mention only is normal
    - What about capitalisation of common names – all common names, or only those with proper nouns?
2. Use of terms to describe the site, and capitalisation of these:
    - Site?
    - site?
    - Application Site?
    - or the specific name of the site?
3. Use of terms to describe the project:
    - the Development?
    - the proposed development?
    - the Scheme?
    - the project?
    - or a specific name of project?

4. Measurements used to describe distance should generally use consistent units. For example, try to avoid:

   SSSI A is 570m and SSSI B is 1.8km from the site boundaries.

There are also some regularly occurring typos. There are two that seem to occur pretty much constantly:

1. Licence/license

   - In British English, 'licence' is the correct spelling for the noun, e.g. *a licence will be required*. License would be incorrect in this usage (although is correct in American English).

   - In British English, 'license' is the correct spelling for the verb, e.g. *the works will need to be licensed*, or *an application will need to be submitted to the licensing department*.

2. Affect/effect

   'Affect' is a verb, so you can affect something. 'Effect' is a noun, so you can have an effect.

   > *The weather conditions are likely to have affected the outcomes of the survey.*
   >
   > *It is likely that the cool weather at the time of the survey had an effect on the behaviour of the target species.*

Chapter 17

Tips for those reviewing reports

Many ecologists will find themselves in a position where they need to do a critical review of a report produced by another ecologist. By this I mean a review by an external organisation – not a review by someone working for the same company or organisation that produced the report (which I've described in the previous chapter as a 'technical review').

This critical review might be undertaken by someone acting in a number of different capacities, such as:

1. An Ecologist or Biodiversity Officer employed by a Local Planning Authority, reviewing ecology reports submitted by the developer as part of a planning application;

2. A client (or an ecologist employed or contracted by a client) who has commissioned an ecology study;

3. An ecologist not involved with the production of a particular ecological report, but who is looking to gain information from it that could be of relevance to a new study (for the same site or a nearby site, for example);

4. An ecologist not involved with the production of a particular ecological report and who is representing a developer promoting a rival site to the site that the report relates to;

5. An interested third party, who may wish to comment on a planning application.

A critical review means an assessment of whether the report is fit for purpose. The conclusions of the report should not be simply accepted but challenged. Is there sufficient data to back up the conclusions, and have the data been collected in a robust manner? Are any opinions suitably justified? Can the data and conclusions be relied upon?

So, if you find yourself in the position of having to undertake a critical review of an ecology report, how do you do it and what specifically should you be looking for?

There will be a whole series of different approaches to this, all of which might be equally valid. Personally, I tend to read it from start to finish, following all cross-references as required, but then going back to the point I'd got to, whilst keeping those 10 Key Characteristics in mind.

Firstly, look at the *Title Page*, any Quality Assurance details, and the *Contents Page*. You should be able to answer these questions:

- Who has written the report?
- Are they **competent**?
- When was the report written and is it still valid?
- Which version of the report are you looking at and is it the latest version (you might not be able to answer this in all cases)?
- What level of Quality Assurance (if any) has been done, who did it and are they **competent**?
- Do the headings and sub-headings in the *Contents Page* accord with what you would have expected to see? Are they similar to most other reports of that type, or as per CIEEM's guidelines on report writing, or as per those suggested in this book?
- Does the report appear to be **well structured**?

Next, read the *Introduction* carefully. This might provide answers to some of the questions asked already, but you should also be able to answer the following questions:

- What is the specific purpose (or purposes) of the report?
- Is that the correct purpose (or purposes) for that report?
- Who is the intended **target** audience for the report?
- If a report isn't clear about its purpose(s), then there's a good chance that it isn't going to be fit for its intended purpose.
- Where is the site? Are its location and boundaries accurately described and shown on appropriately scaled drawings?
- If the report has been produced in relation to a specific development project, what does that project entail (in basic terms at least)?

You will need to bear in mind, dependent upon the capacity you're acting in, that the report might have been written for a different purpose from the one you intend to use it for.

Next, for most ecological reports, you will need to read the *Methods* and *Results* sections, as well as any associated appendices. You will need to check cross-references to figures or drawings and photographs thoroughly. There are some further questions you should be able to answer as you do this:

- Is the report **transparent and truthful** about the way that data have been collected, including the methods used and the sources of information?
- Is the desk study information **robust**?
  - Have data been collected over a sufficiently large study area?
  - Have all relevant sources been interrogated?
  - When were the data collected, and are they still valid?
- Is the field survey information **robust**?
  - Have data been collected over a sufficiently large study area?
  - Have appropriate survey techniques been used?
  - Were the surveyors suitably competent and (where appropriate) licensed?
  - Do the survey methods accord with relevant good practice guidelines?
    Note – I've seen many reports containing a statement that the methods accord with such guidelines, when in reality they don't, so to answer

this question you need to look beyond such statements and actually compare the methods with the specific guidelines referred to.
  - Are any departures from guidance clearly explained and **justified**?
  - When were the data collected, and are they still valid?
  - Have all relevant data been presented in the report, including details of dates, times and weather conditions during surveys, where relevant?
  - Is the presentation of data sufficiently **clear and precise** to allow the report to fulfil its purpose(s)?

- Have all possible limitations (distinguished from 'constraints' – see Chapter 8 for further details) to the collection of information been identified and discussed, with their implications clearly explained?

Of course, some ecological reports won't have *Methods* and *Results* sections, such as a Habitat Management Plan. However, any report will be based on some form of ecological data, and you will therefore need to amend this list of questions to address the same issues accordingly.

Your next task, in most cases, is to read and review the 'interpretation' sections. This might include a *Discussion, Assessment, Recommendations, Impact Assessment* or *Mitigation* section, for example. The list of questions to be asked here will therefore vary depending on the type of report you're reviewing, and its intended purpose. However, the following questions will be relevant in many cases:

- Have opinions (such as assessment of value, likely presence/absence, or adequacy of mitigation measures) been described in a way that makes it **clear** that they are opinions rather than facts?

- Have all opinions been **justified**, such as by according with good practice guidelines or the outcomes of relevant research?

- Are the assertions made supported by the data presented in the *Results* section (or equivalent)?

- If the answer to the previous question is no, is that because the data have been misinterpreted, or because the dataset is incomplete?

- Look out for sweeping statements with no justification, or generalisations.

- Is any assessment of impacts specific and **precise**, adequately expressing the scale and duration of any impacts (using numbers in a specific manner, rather than vague terms like 'small' or 'minor' for example)?

- Do any mitigation or compensation measures proposed accord with relevant good practice guidelines?

- Does the report clearly distinguish between different 'levels' of constraint (distinguished from 'limitations' – see Chapter 8 for further details), dealing with them in a **proportionate** manner?

- Have opinions been expressed in an **impartial** manner – are they 'balanced'?

You will then need to read the *Conclusions* section and check for a few things in particular:

- Are the conclusions **justified** and supported by the text presented in the report?

- Do they follow logically from the information presented?

- Make sure that nothing is claimed in the *Conclusions* section that isn't mentioned earlier in the report, or doesn't follow logically from what has been presented.

- Do the conclusions relate to the purpose(s) of the report?

It is always worth checking any appendices, figures or photos in the back of the report to make sure that there's nothing unexpected lurking there. In reality, you should have looked at all of this already as, assuming that it's relevant, you will have been directed to it by following a cross-reference from an earlier part of the report. There shouldn't be anything in the appendices, or any figures or photos, that aren't cross-referenced in the report. Nevertheless, I'd suggest having a quick flick through the back of the report to make sure that's the case. Ignore anything generic that's been included and is not specifically relevant – just look for any specific information provided that has (presumably accidentally) not been referred to. Make sure this is all as you would expect, and doesn't change any of your answers to the questions set when reading the earlier sections.

Finally, read the *Summary*. If you're part of the intended target audience then you have a right to ensure that the *Summary* is well written and helpful to you. You can check this against the summary provided at the end of Chapter 13. If you're not part of the intended target audience then whether the *Summary* is well written or not isn't really your concern. However, in most cases, you will still need to check that the *Summary* isn't misleading, bearing in mind that some people might only read the *Summary*. So, read it and make sure that it accurately reflects what you've read in the rest of the report.

There are, of course, some overlaps in terms of what to look out for between 'technical reviews', as described in the previous chapter, and 'critical reviews'. Understanding what the client or Local Planning Authority's Ecologist is looking for when they read the report is obviously worth thinking about when a draft report is being read by a senior colleague before signing it off.

Some types of report will have their own specific set of rules to follow, which will provide useful guidance for someone doing a critical review. For Ecological Impact Assessment (EcIA) Reports, this is where the EcIA Checklist that I mentioned in Chapter 16 comes in. This was published by CIEEM in 2019. It effectively provides a starting point for a standard list for those undertaking critical reviews (or technical reviews – see Chapter 16) of EcIA Reports. It provides a series of criteria to be used in judging whether an EcIA Report is fit for purpose.

Chapter 18

Referencing sources

The importance of referencing sources

Throughout this book I've discussed the importance of using appropriate sources to justify approaches taken in a report. This is something that those with any sort of scientific background will already be familiar with. However, professional ecological reports tend to contain fewer reference sources than a piece of scientific literature published in a peer-reviewed journal, for example. There are at least a couple of reasons for this.

Firstly, in many cases, professional ecological reports are written for a purpose that has a single overarching good practice guidance document. This isn't always the case, but where such a document exists it will often be the only necessary reference source for a given species or habitat, in the context of a specific report.

Secondly, professional ecological reports have to be written in a style that is accessible to the non-ecologist. As a result, the authors tend not to qualify every single statement with a reference, as you might find in a scientific paper. Statements that are likely to be widely accepted and uncontroversial are therefore often not supported with a reference. This, though, is a tricky balancing act, as what is considered to be widely accepted and uncontroversial by one person may not be by another. Also, this practice can lead to the general acceptance of reports that routinely provide few or no reference sources as evidence, which should be actively discouraged.

If in doubt then, provide a reference source to justify or qualify a statement.

There are several places in a report where a reference source will be needed:

1. In the *Introduction*. Any background information, such as previous survey reports or overarching guidance documents, will need to be appropriately referenced.

2. In the description of desk study methods. Desk studies will normally include a search for information using online resources as well as a search with a Local Environmental Records Centre (or equivalent). Any online resources used will need to be referenced.

3. In the description of survey or assessment methods, such as for a detailed description of a survey technique, or for justification of the selection of a certain level of survey effort. This might be a reference to a specific guidance document that sets out the recommended approach to surveying for a given habitat or species. Or, where such a document doesn't exist, a survey protocol may have been designed based on an adapted version of another survey method, to make it appropriate for the intended purpose of the report. Also, in some cases there will be multiple possible approaches, and the reasons for selecting a specific approach will need to be **justified**. This will require reference to the appropriate sources.

4. In the assessment of value or importance (where relevant). Ecological Impact Assessment (EcIA) Reports, for example, will normally assign a geographic scale of importance or value to each relevant ecological feature (a designated nature conservation site, a habitat or a species population). This can only really be done through the use of appropriate sources of contextual information, such as the local Biodiversity Action Plan, County Mammal Atlas or County Bird Report, for example. Any contextual information being used to underpin the assessment should be appropriately referenced.

5. To justify an assessment of potential impacts on any given ecological feature. EcIA Reports and, to a certain extent, PEA Reports will need to make an assessment of how a designated site, a habitat or a species population could be affected by a specific development project. This will require the author to effectively provide an opinion, which should

be based on available evidence (or a **clear** statement made that there is no evidence). This evidence should be referenced. Let's take an example. The author of an EcIA Report has to assess the impact of a proposed housing development on Species X. They think that Species X is unlikely to be materially affected, because the only mechanism for an impact on Species X is through disturbance due to construction noise, and Species X is not particularly susceptible to disturbance from noise. Assuming that the report provides sufficient detail to allow the reader to agree that impacts from the development besides disturbance due to construction noise are unlikely, the assessment hinges on the statement that Species X is not particularly susceptible to this specific impact. This sort of statement would need to be backed up with a reliable reference source, such as an approved guidance document or peer-reviewed research papers.

6. To justify an approach to mitigation or compensation. Any description of a proposed mitigation or compensation measure should explain why a particular approach is proposed, and what the evidence is that it is likely to be effective. Once again, good practice guidance documents will often provide recommendations for mitigation or compensation, and should be referenced as necessary to highlight that a recommended approach is being followed. Evidence may also be found in papers published in peer-reviewed journals, other published articles providing opinions or case studies from professional ecologists (such as might be found in CIEEM's *In Practice* magazine), web-based sources (such as Conservation Evidence), published research reports or (often unpublished) monitoring reports.

Citing reference sources in the text

The accepted approach is to provide the author's surname and year of publication in brackets at the end of the sentence or the end of the relevant clause in some cases. I've illustrated this using some, clearly fictional, examples.

Where there are two authors, both surnames should be provided, such as:

> *A recent study has shown that unicorns are always found within 200m of the end of a rainbow (Smith and Jones 2019).*

Where there are three or more authors it is normal to include the first surname followed by '*et al.*' (in italics), such as:

> *A previous study had recorded only pots of gold at the end of rainbows (Jones et al. 2017).*

I've put this entire example into italics, but note that the reference should be written '… (Jones *et al.* 2017)' with '*et al.*' in italics.

Published documents where the author is an organisation should cite the name of the organisation and year of publication.

And where there are multiple sources being referred to, the authors and dates should be provided in the same set of brackets, subdivided by semicolons. For example:

> *Studies from 2018 onwards have shown that unicorns are always found within 200m of the end of a rainbow (Jones 2018; Smith and Jones 2019; Smith 2020).*

It is fairly unusual to see multiple sources being referred to in a professional ecological report for two main reasons:

1. Providing references for widely accepted statements, and which multiple sources therefore refer to, is generally unnecessary;[1] and

2. Because there are relatively few relevant reference sources, and where there are multiple sources there is often one that effectively 'trumps' the others, meaning that only one needs to be referred to.

To illustrate this I'll need to move away from fictional examples. If I want to make a statement relating to something that really is widely accepted, then I don't necessarily need to reference a source. For example:

> *Badgers live in setts. A single social group of badgers can occupy multiple different setts, each with a different purpose and different frequencies of use. The term 'main' sett is used to refer to a sett that is in constant use by*

---

1 I appreciate the irony of making this point in relation to the previous statement about unicorns.

> *badgers (other types of sett will not necessarily be continually occupied), and is generally used for breeding (Cresswell et al. 1990).*[2]

The first two sentences don't really need a reference source in the context of a professional ecology report as they are widely accepted. The third sentence refers to a specific definition of a term, and therefore does. The definition is used in multiple references, but only one is really needed as there is agreement on the points being made – it should ideally be the original source which is referenced, unless there is a more recent, widely accepted source that is more appropriate in a given case.

The Harvard system

The full reference source needs to be provided, and this will normally be done using the Harvard system.

For books or published guidance document, this means:

> Author's surname, Author's initials (year of publication) *Title of book written it italics.* Name of publisher, City of publication.

Where there are multiple authors these should all be listed, separated by commas. Where the author is an organisation, this should be provided.

For example:

> CIEEM (2017) *Guidelines on Ecological Report Writing, 2nd Edition.* Chartered Institute of Ecology and Environmental Management, Winchester.

> Dean, M., Strachan, R., Gow, D. and Andrews, R. (2016) *The Water Vole Mitigation Handbook (Mammal Society Mitigation Guidance Series).* Eds Fiona Mathews and Paul Chanin. Mammal Society, London.

---

2 Cresswell, P., Harris, S. and Jefferies, D.J. (1990) *The History, Distribution, Status and Habitat Requirements of the Badger in Britain.* Nature Conservancy Council, Peterborough.

For journal articles, this means:

> Author A surname, Author A initials, Author B surname, Author B initials, etc. (year of publication) Title of article. *Title of publication written it italics* Volume number, Page numbers.

For example:

> Wansbury, C., Linford, J., Lawrie, V., Atherton, L., Barwig, J. and Pugh, C. (2014) Communication skills for ecologists – to influence policy on biodiversity and ecosystem services we must know our audiences. *In Practice* 86, 15–19.

Referencing web sources

For citations of web sources it is a little more complex, as this might relate to articles published online or it might relate to accessing a website used to search for information, such as the NBN Atlas[3] or MAGIC.[4]

Where the reference is to an online article, the reference should be provided as described for books and journals as far as possible, with the addition of the web address and the date that the website was accessed.

Where the reference is to a specific website used to search for information, it is not normally necessary to provide a full reference source in the document, but the text should include the web address and the date that the website was accessed. For example:

> The MAGIC website (https://magic.defra.gov.uk) was used to search for statutory designated nature conservation sites within 2km of the proposed development. The website was accessed on 30 June 2019.

---

3 https://nbnatlas.org

4 https://magic.defra.gov.uk

Footnotes, endnotes or a References section

Providing full reference source can be done in two basic ways:

1. In a *References* section – listing all the references used, in alphabetical order at the end of the report; and

2. As a footnote – including a reference number in the main text and providing the full reference at the bottom of the page.

References sections are the traditional way of dealing with this. Increasingly, footnotes are being used as an alternative.

Footnotes have the obvious advantage of providing the full reference source on the same page as the text that refers to it. This can work very well. However, it can cause problems in two particular scenarios. Firstly, where there are multiple reference sources on the same page you can end up with very little actual text on a page, but lots of footnotes. This looks untidy and doesn't read well. Secondly, where a single reference source is referred to multiple times you either have to add multiple number references back to the original reference, or repeat it. In either of these cases I'd argue that the traditional approach of a *References* section works better.

I have also seen references provided using endnotes – including a reference number in the main text and providing the full reference at the end of the document. This is different from a *References* section in that the references are numbered sequentially and therefore appear in the order they are referred to in the document, rather than in alphabetical order. Personally I would avoid using endnotes, as they can suffer from the second problem that I highlighted for footnotes and there's no significant advantage of providing references in number order rather than alphabetical order. I would therefore use a *References* section in preference over endnotes. Endnotes tend to work better for supporting text that couldn't be effectively listed in alphabetical order.

Bibliography or References section?

A *Bibliography* is different from a *References* section in a report. A *References* section lists all of the sources referred to in the report, whereas a *Bibliography* section lists sources of information that may relate to the topic of the report, but aren't necessarily directly referenced – a suggested further reading list for those sufficiently interested. It isn't necessary to include a *Bibliography* section in most professional ecological reports, but a *References* section is needed (unless the references are provided as footnotes, as discussed already).

Chapter 19
How long is a report valid?

The vast majority of the reports produced by professional ecologists will have a finite lifespan. This is because they relate to the natural world and therefore will, of course, change over time.

Most of the habitats that we will be writing about are the result of management, to some extent at least. They could therefore change into a different habitat if the management changes or ceases. And the species present on a site will be dictated largely by the habitats present and the natural range of that species. So if the habitats change then the species likely to be present will change, and if the natural range of a species increases or decreases this will also affect the chance of encountering them. Of course there are a range of other factors at play; some of those might also change, such as disturbance through human use.

We therefore have to accept that an ecological report will become out-of-date at a certain point, as will the survey data that it is based upon. It could be argued that this happens as soon as the report is written, or as soon as the final survey visit is completed. There are no guarantees that site conditions will not have changed. However, this isn't very helpful for reasons that are hopefully fairly obvious. As professional ecologists, and particularly ecological consultants, we need to be careful to avoid being too precautionary. We are, after all, the ones that will benefit if a report, and the data that it is based on, becomes obsolete and need updating – that's more work for us, which we will expect to be paid for.

The key to dealing with this issue in a balanced way is, firstly, to try to 'future proof' our reports as far as possible. We can do this by considering:

1. The likelihood of habitat changes occurring. Some of the habitats being considered will be unlikely to change in the short term. Others will be much more likely to change at some point in the future, and there will even be a considerable degree of certainty about when that might occur in some cases.

    - A commercially managed plantation woodland for example – areas of woodland may well be clear-felled when they reach a certain age and then be replanted.

    - An arable field will be sown with a particular crop in one year but might be earmarked for a different crop next year; although we might still call this arable farmland, the species that it supports will change.

    - A patch of bare ground might become covered in weeds and then eventually develop into bramble-dominated scrub.

2. The likelihood of the natural range of a species changing

    Again, this will vary for different species and won't be the same in all locations. Having an understanding of the status of a species in a given area is therefore vital. Is it expanding? Is it contracting? Are there any initiatives under way that might affect this? Contextual information from a desk study will help with this, as will historical records, but they won't cover this entirely. Knowing that landscape-scale mink control is about to be introduced in a given location is likely to have a bearing on your opinion as to the likelihood of future expansion of the water vole population, for example. Local knowledge is therefore critical.

3. The reliability of the survey data

    Some survey techniques are more reliable than others. How many sampling visits have been undertaken? Is it a mobile species that may not have been present at the time of the surveys? Or a species that may only utilise the habitats present at certain times of year? Were there any survey limitations?

4. Whether the presence of a species was ruled out because the site did not provide valuable habitat, or because a survey confirmed likely absence.

   Clearly, if a site supports suitable habitat for a species then it is more likely that they may move into it (assuming they are present in the local area) than if the site doesn't support suitable habitat – unless it's likely that the habitat changes.

5. Are there any other relevant factors that are dictating the presence or (likely) absence of a particular habitat or species?

Giving some thought to how ecological features might change in the future is a key part of the Ecological Impact Assessment process, so this idea of 'future proofing' won't be new to anyone used to writing EcIA Reports. But there's no reason why it shouldn't also be considered in other types of ecological report – standard survey reports or PEA Reports, for example.

And the extent to which you are able to reliably 'future proof' a report will provide a useful indication of its likely lifespan and the assumptions that any such assessment is based upon.

Secondly, we should be **clear** about the likely longevity of a report, or the survey data, in the report itself. It's good practice for an author to specify a time frame for the validity of their report, and the survey data it contains. They are likely to be the best-placed person to do this, as they will have the best understanding of the factors described above and the implications of them in the context of a specific report.

Thirdly, once we have a clear picture of the factors that might dictate that a report, or the survey data in it, is no longer valid, we can make informed decisions about this rather than having arbitrary cut-offs. Those decisions would still need to be made by an ecologist, and a site visit or revision of desk study might well still be appropriate in order to allow a robust decision to be made.

CIEEM published an advice note on this issue in April 2019.[1] It sets out some broad advice on likely time frames, as well as giving example scenarios.

---

1 CIEEM (2019) *Advice Note on the Lifespan of Ecological Reports & Surveys*. Chartered Institute of Ecology and Environmental Management, Winchester.

In general terms, CIEEM's advice note suggests that a report or survey that is less than 12 months old is likely to be valid in most cases. Obviously there will be circumstances where this isn't the case, but report authors really need to make sure that their report is 'future proofed' to this extent at least. And where there are identifiable reasons to suggest otherwise these must be clearly stated.

The advice note also states that reports or surveys in the region of 12–18 months old are also likely to be valid in most cases, but gives some specific examples of when this might not the case. Again, I'd suggest that report authors need to try to 'future proof' the report to this extent at least.

The risk of a report or survey being out-of-date increases significantly when they become more than 18 months old. It may be possible to 'future proof' a report for longer than 18 months, but this will become increasingly difficult. The CIEEM advice note therefore recommends that the validity of reports or surveys that are between 18 months and three years old needs to be reassessed by a professional ecologist. They will consider the factors described above, and may find that surveys don't need to be repeated as their results are likely to hold true; however, this can't be assumed.

Beyond three years old the same principle applies, but the likelihood of surveys needing to be updated increases.

In summary then, report authors need to:

1. Do their best to 'future proof' a report as far as is reasonably possible, with 18 months as a realistic minimum in most cases. However, care needs to be taken not to either overestimate or under-estimate in relation to this.

2. Make a **clear** statement in relation to the likely longevity of a report and its survey data. This will be particularly important where there is a reasonable justification for the report becoming out-of-date within 18 months.

3. Give some explanation of the factors that have been considered and any assumptions made. This should allow a reassessment to be made over the continued validity of the report at any point in the future.

Chapter 20

Useful sources of information

There are a number of guidance documents and other relevant information sources that you will want to be familiar with, and have to hand when you're writing ecological reports. I've mentioned some of them in the previous chapters of this book, but have also provided a handy list here. Some of these relate specifically to certain types of report, but may well be more widely applicable as well. This isn't an exhaustive list – there will be many others that I haven't listed here that will also be useful.

British Standards Institution (2013) *BS42020:2013 Biodiversity – Code of practice for planning and development.* BSI Standards Limited.
The British Standard on Biodiversity sets out recommendations to ensure that ecological information used within the planning process is of an appropriately high quality. It includes information on the structure and content of ecological reports, and is therefore directly relevant to the subject of this book. It also contains sections relating to professional competence, professional judgement, identifying limitations, full disclosure of scientific method, and the appropriate use of planning conditions. These, along with many other sections, are all relevant to the preparation of ecological reports that are going to be used in some stage of the planning process.

CIEEM (2017) *Guidelines on Ecological Report Writing, 2nd Edition.* Chartered Institute of Ecology and Environmental Management, Winchester.
This provides some specific guidance on structure and content of ecological reports, particularly survey reports, PEA Reports and EcIA Reports. It is one of the key reference sources to be used by professional ecologists when determining the structure and content of reports. It can be downloaded from the CIEEM website.

CIEEM (2017) *Guidelines for Preliminary Ecological Appraisal, 2nd Edition*. Chartered Institute of Ecology and Environmental Management, Winchester.
This provides advice in relation to undertaking a PEA, but includes advice on when a PEA Report can be submitted to accompany a planning application, rather than an EcIA Report. It can be downloaded from the CIEEM website.

CIEEM (2017) *Guide to Ecological Surveys and Their Purpose*. Chartered Institute of Ecology and Environmental Management, Winchester.
This document defines some of the commonly used terms for ecological surveys and names for ecological reports, in an attempt to promote standardised terminology. It can be downloaded from the CIEEM website.

CIEEM (2018) *Guidelines for Ecological Impact Assessment in the UK and Ireland: Terrestrial, Freshwater, Coastal and Marine*. Chartered Institute of Ecology and Environmental Management, Winchester. Version 1.1, updated September 2019.
This document sets out the approach to undertaking the Ecological Impact Assessment process. It also provides a skeleton 'template' for an EcIA Report, which follows that provided in the *Guidelines on Ecological Report Writing*. It can be downloaded from the CIEEM website.

CIEEM (2019) *Advice Note on the Lifespan of Ecological Reports & Surveys*. Chartered Institute of Ecology and Environmental Management, Winchester.
This advice note provides a benchmark for the length of time that ecological surveys and reports should be considered valid for, although it highlights that surveys and results may be valid for longer periods in some cases (which requires corroboration) or possibly shorter periods in some cases (which should be clearly stated in a report, where it is the case). It can be downloaded from the CIEEM website.

CIEEM/ALGE (2019) *Ecological Impact Assessment Checklist*. Chartered Institute of Ecology and Environmental Management, Winchester and the Association of Local Government Ecologists.
This document provides a checklist intended for use in determining whether an EcIA Report is fit for purpose or not.

CIEEM Professional Standards Committee (2016) Pragmatism, Proportionality and Professional Judgement. *In Practice* 91: 57–60.
This article appeared in CIEEM's quarterly publication *In Practice*. It relates to the use of professional judgement – when is it appropriate to use

professional judgement and how is the term defined? This issue is relevant to all aspects of a professional ecologist's work but has particular relevance to how opinions are expressed in reports. CIEEM members can download the article from the CIEEM website.

DTA Publications. The Habitats Regulations Assessment Handbook (https://www.dtapublications.co.uk)
This online handbook provides advice on all aspects of Habitats Regulations Assessment, including how to present the outcomes of different stages.

Freeths LLP (2020) Copyright Considerations in Ecological Reports. *In Practice* 108: 55–56.
This article also appeared in CIEEM's quarterly publication *In Practice*. It deals with the issue of copyright in relation to professional ecological reports. Effectively – who 'owns' the report. I've not provided a chapter on this topic in the book, as I see it as an area best left to those in the legal profession to advise on. CIEEM members can download the article from the CIEEM website.

Some of the more recent species-specific guidance documents produced for professional ecologists also contains recommendations on how surveys reports should be set out, in relation to a particular species (or group of species). For example:

Collins, J. (ed) (2016) *Bat Surveys for Professional Ecologists: Good Practice Guidelines, 3rd Edition*. The Bat Conservation Trust, London.
This document sets out recommendations for undertaking a range of different types of bat survey. It also includes an entire chapter on 'writing bat reports', providing a skeleton template for them.

Dean, M., Strachan, R., Gow, D. and Andrews, R. (2016) *The Water Vole Mitigation Handbook (Mammal Society Mitigation Guidance Series)*, edited by Fiona Mathews and Paul Chanin. Mammal Society, London.
This document sets out guidelines for carrying out water vole surveys and the design and implementation of mitigation measures for the species. Although it doesn't provide a skeleton template for water vole related reports, it provides advice on the contents of survey reports and mitigation strategies.

Appendix A: Suggested headings and sub-headings for an ecological survey report

Based on, and adapted from, CIEEM's *Guidelines on Ecological Report Writing* (CIEEM 2017).

Heading/sub-heading		Content
	Title Page	
	Contents Page	
	Summary	• A summary of the salient points
1.	Introduction	• Purpose of the report • Site location • Brief description of the habitats present within the survey site as well as some contextual information for the surrounding area • Details of the proposed project (assuming that is the reason for the survey being undertaken) • Details of who has produced the report, their competence and who they were contracted by • Any other relevant background information
2.	Methods	
2.1	Desk study	• A description of the desk study methods, including sources, information searched for/requested, date of search, search area
2.2	Field survey	• A description of the field survey methods used, including dates of survey(s), times of survey(s) (where relevant), names of surveyor(s), weather conditions, study area (with reference to a map, ideally), and details of the survey method • An explanation of how the survey methods accord with relevant good practice guidelines (or justification for any departure from such guidance) • The field survey methods need to be described in sufficient detail to allow someone else to repeat the survey and obtain a comparable result • Detailed survey methods can be moved to an appendix to reduce the length of this section, if appropriate
2.3	Limitations	• A description of any limitations associated with the collection of data[1]

1 Note that the implications of any limitations must also be explained – either in the Limitations section or later in the report.

Heading/sub-heading		Content
3.	Results	
3.1	Desk study	• A description of relevant information collected as part of the desk study
3.2	Field survey	• A description of the field survey results • Detailed survey results can be moved to an appendix to reduce the length of this section, if appropriate • Photographs can be provided here, or moved to an appendix
4.	Discussion[2]	• A discussion of the implications of the results in the context of the purpose of the study • An explanation of the implications of any limitations to the survey
5.	Conclusions and recommendations	• Conclusion(s) in the context of the purpose(s) of the study – have the aims been achieved? • Appropriate recommendations to deliver the purpose of the study (if it hasn't been achieved) • Appropriate recommendations for further work (assuming that this is within the scope of work)
6.	References	• A list of any documents referred to in the text (or these can be provided as footnotes)
	Figures	• Figures to illustrate the methods and results (where appropriate)
	Appendices	• Detailed survey methods (if appropriate) • Detailed survey results (if appropriate) • Photographs (if appropriate)

2 Note that this section can be combined with the Results in some cases to help the report flow more logically and reduce unnecessary repetition.

Appendix B: Suggested headings and sub-headings for Preliminary Ecological Appraisal Reports

Based on, and adapted from, CIEEM's *Guidelines on Ecological Report Writing* (CIEEM 2017).

Heading/sub-heading		Content
	Title Page	
	Contents Page	
	Summary	• A summary of the salient points
1.	Introduction	• Purpose of the report • Site location • Brief description of the habitats present within the survey site as well as some contextual information for the surrounding area • Details of the proposed project • Details of who has produced the report, their competence and who they were contracted by • Any other relevant background information
2.	Planning policy and legislation	• A summary of the relevant planning policy (national and local) and relevant legislation • Should be project specific
3.	Methods	
3.1	Scope of study	• An explanation of the basis for the approach taken to data collection • A description of the likely zone(s) of influence of the project – which will have informed study areas • Reference to any consultation undertaken/pre-app advice received
3.2	Desk study	• A description of the desk study methods, including sources, information searched for/requested, date of search, search area
3.3	Field survey	• A description of the field survey methods used, including dates of survey(s), times of survey(s) (where relevant), names of surveyor(s), weather conditions, study area (with reference to a map, ideally), and details of the survey method • An explanation of how the survey methods accord with relevant good practice guidelines (or justification for any departure from such guidance) • The field survey methods need to be described in sufficient detail to allow someone else to repeat the survey and obtain a comparable result • If any detailed surveys (beyond a walkover type survey) have been undertaken, then detailed methods can be moved to an appendix to reduce the length of this section, if appropriate

Heading/sub-heading		Content
3.4	Limitations	• A description of any limitations associated with the collection of data[1]
4.	Results	
4.1	Designated sites	• Information relating to designated nature conservation sites within the study area, including name and description of site, confirmation of type of designation, location relative to the site (on site/off site, distance and direction from site), reasons for designation
4.2	Habitats	• A description of the habitats present within the study area, with reference to a suitable map/figure and with appropriate photographs (which can be moved to an appendix) • The focus should be on habitat types that are relevant to the assessment (i.e. in England, Scotland and Wales this would mean focusing on Habitats of Principal Importance For the Conservation of Biodiversity, habitats listed as priorities for conservation in the local Biodiversity Action Plan, and any protected habitats).
4.3 et seq	Relevant species and/or species groups, e.g. • Plants • Invertebrates • Amphibians • Bats • Etc.	• A description of the results of the various elements of the study for each species or group of species, focusing on protected species, Species of Principal Importance for the Conservation of Biodiversity, species listed as priorities for conservation in the local Biodiversity Action Plan, and invasive non-native species. • This should combine the results of the desk study and the walkover survey, as well as any other more detailed surveys undertaken. The source of the information presented should be made clear – this can be done using sub-headings for example.
5.	Ecological constraints and recommendations for mitigation	• Identify the ecological constraints to the proposed project • Provide advice on measures to avoid potentially significant effects, and/or design changes, as appropriate • Set out the principles of any mitigation and licences likely to be required (as far as possible at this stage)
6.	Recommendations for further surveys	• Identify the further surveys likely to be required to inform an ecological impact assessment for the proposed project
7.	Recommendations for enhancement/ Biodiversity benefit/Biodiversity Net Gain	• Identify appropriate opportunities to deliver required benefits for biodiversity (the specific requirements and terminology used will be different in different countries)
8.	Conclusions	• Conclusion(s) in the context of the purpose(s) of the study
9.	References	• A list of any documents referred to in the text (or these can be provided as footnotes)
	Figures	• Appropriate figures to illustrate the results
	Appendices	• Detailed survey methods (if appropriate) • Detailed survey results (if appropriate) • Photographs

1 Note that the implications of any limitations must also be explained – either in the Limitations section or later in the report.

Appendix C: Suggested headings and sub-headings for Ecological Impact Assessment Reports

Based on, and adapted from, CIEEM's *Guidelines on Ecological Report Writing* (CIEEM 2017).

Heading/sub-heading		Content
	Title Page	
	Contents Page	
	Summary	• A summary of the salient points
1.	Introduction	• Purpose of the report • Site location • Brief description of the habitats present within the survey site as well as some contextual information for the surrounding area • Details of the proposed project • Details of who has produced the report, their competence and who they were contracted by • Any other relevant background information
2.	Planning policy and legislation	• A summary of the relevant planning policy (national and local) and relevant legislation • Should be project specific
3.	Methods	
3.1	Scope of study	• An explanation of the basis for the approach taken to data collection • A description of the likely zone(s) of influence of the project – which will have informed study areas • Reference to any consultation undertaken/pre-app advice received
3.2	Desk study	• A description of the desk study methods, including sources, information searched for/requested, date of search, search area

Heading/sub-heading		Content
3.3	Field survey	• A brief description of the field survey methods used • An explanation of how the survey methods accord with relevant good practice guidelines (or justification for any departure from such guidance) • Detailed methods, including dates of survey(s), times of survey(s) (where relevant), names of surveyor(s), weather conditions, study area (with reference to a map, ideally), and details of the survey method should be provided in an appendix to reduce the length of this section. The field survey methods need to be described in sufficient detail to allow someone else to repeat the survey and obtain a comparable result
3.4	Limitations	• A description of any limitations associated with the collection of data[1]
3.5	Assessment methods	• A description of the methods used for the assessment of impacts • Identification of the scales of importance being used • Description of how a 'significant effect' is being defined
4.	Baseline conditions	
4.1	Selection of Important Ecological Features	• List of 'Important Ecological Features' to be taken forward as part of the assessment and justification for 'scoping out' features not being taken forward
4.2 et seq	Sub-heading for each Important Ecological Feature	• A description of the baseline conditions for each Important Ecological Feature, taking into account the results of the field surveys and desk study, and any other relevant information about habitat condition or management, and any possible future changes to the baseline. • Assessment of the scale of importance of each Important Ecological Feature, with appropriate justification
5.	Scheme description	• A description of the proposed scheme, with reference to appropriate drawings • Include changes to scheme design that have been implemented to avoid/minimise impacts on biodiversity

// ctd.

1 Note that the implications of any limitations must also be explained – either in the Limitations section or later in the report.

Heading/sub-heading		Content
6.	Potential impacts and proposed mitigation measures	
6.1	Project-wide mitigation	• Set out the ecological mitigation measures that apply to multiple Important Ecological Features, such as: – Production of a Landscape and Ecological Management Plan and subsequent management of retained habitats in accordance with that plan – Production and implementation of a Construction Environmental Management Plan, including appropriate pollution control measures during construction – Installation (and subsequent maintenance) of tree protection fencing prior to construction works commencing to protect retained trees and hedgerows, in accordance with the relevant Arboricultural Impact Assessment
6.2 et seq	Sub-heading for each Important Ecological Feature	• Identify and characterise the potential impacts of each phase of the project on each Important Ecological Feature • Identify and describe the mitigation measures proposed for each Important Ecological Feature • Assess the significance of residual effects for each Important Ecological Feature
6.x	Summary of mitigation measures	• Provide a summary table of mitigation measures proposed
7.	Compensation	• Describe any compensation proposed to offset significant residual effects
8.	Cumulative effects	• Identify any other projects that could result in cumulative effects • Provide an assessment of the likelihood of significant cumulative effects for each Important Ecological Feature assessed.
9.	Enhancement/ Biodiversity benefit/ Biodiversity Net Gain	• Identify appropriate proposals to deliver required benefits for biodiversity (the specific requirements and terminology used will be different in different countries)
10.	Monitoring	• Outline proposals for any ecological monitoring required to confirm the effectiveness of mitigation, compensation or enhancement measures
11.	Conclusions	• Conclusion(s) in the context of the purpose(s) of the study
12.	References	• A list of any documents referred to in the text (or these can be provided as footnotes)
	Figures	• Appropriate figures to illustrate the results of surveys and impacts and mitigation, as necessary
	Appendices	• Detailed survey methods • Detailed survey results • Photographs

Appendix D: Suggested headings and sub-headings for Biodiversity Action Plans or Strategies

Heading/sub-heading		Content
	Title Page	
	Contents Page	
	Summary	• A summary of the salient points
1.	Introduction	• Purpose of the report • Site location or description of area covered • Brief description of the habitats present within the area covered by the report • Details of who has produced the report, their competence and who they were contracted by • Any other relevant background information
2.	Ecological data	• A summary of the data used to underpin the report, including data sources such as desk study and field survey • In many cases this can cross-refer to other survey reports or published documents, rather than reproducing all of the information in the report • A description of any limitations associated with the collection of data
3.	Biodiversity objectives	• A description of the specific objectives of the Plan or Strategy, and how these relate to national or local conservation priorities, or other relevant documents
4.	Specific measures to protect or enhance biodiversity	• A description of any measures proposed to achieve the objectives • This section will probably be best subdivided by designated sites, habitat or species, which might have their own individual objectives
5.	Additional information required	• Where additional information is needed to inform the Plan or Strategy, this should be described
6.	Monitoring	• Details of monitoring proposed to allow progress against the objectives to be measured
7.	Conclusions	• Conclusion(s) in the context of the purpose(s) of the report
8.	References	• A list of any documents referred to in the text (or these can be provided as footnotes)
	Figures	• Appropriate figures to illustrate the results
	Appendices	• Detailed survey methods (if appropriate) • Detailed survey results (if appropriate) • Habitat or species distribution maps (if appropriate) • Photographs (if appropriate)

Appendix E: Suggested headings and sub-headings for an ecological monitoring report

	Heading/sub-heading	Content
	Title Page	
	Contents Page	
	Summary	• A summary of the salient points
1.	Introduction	• Purpose of the report • Site location • Brief description of the habitats present within the site as well as some contextual information for the surrounding area • Details of the proposed project (assuming that is the reason for the monitoring being undertaken) • Details of who has produced the report, their competence and who they were contracted by • Any other relevant background information
2.	Monitoring methods	• A description of the methods used, including dates of survey(s), times of survey(s) (where relevant), names of surveyor(s), weather conditions, study area (with reference to a map, ideally), and details of the survey method • An explanation of how the survey methods accord with relevant good practice guidelines (or justification for any departure from such guidance) • A description of any limitations associated with the collection of data[1] • This section will be best subdivided by the habitats or species that were the subject of the monitoring
3.	Results	• A description of the results (subdivided by habitat or species) • Detailed survey results can be moved to an appendix to reduce the length of this section, if appropriate • Photographs can be provided, or moved to an appendix
4.	Discussion[2]	• A discussion of the implications of the results in the context of the purpose of the study • An explanation of the implications of any limitations to the survey

1 Note that the implications of any limitations must also be explained – either in the Limitations section or later in the report.

2 Note that this section can be combined with the Results in some cases to help the report flow more logically and reduce unnecessary repetition.

Heading/sub-heading		Content
5.	Conclusions and recommendations	• Conclusion(s) in the context of the purpose(s) of the study – have the aims been achieved? • Appropriate recommendations to deliver the purpose of the study (if it hasn't been achieved) • Appropriate recommendations for changes to monitoring frequency or methods, or any remedial action required
5.	References	• A list of any documents referred to in the text (or these can be provided as footnotes)
	Figures	• Appropriate figures to illustrate the areas for the monitoring
	Appendices	• Detailed survey methods (if appropriate) • Detailed survey results (if appropriate) • Photographs (if appropriate)

Appendix F: Suggested headings and sub-headings for an ecological monitoring strategy

Heading/sub-heading		Content
	Title Page	
	Contents Page	
	Summary	• A summary of the salient points
1.	Introduction	• Purpose of the report • Site location • Brief description of the habitats present within the site as well as some contextual information for the surrounding area • Details of the proposed project (assuming that is the reason for the monitoring being undertaken) • Details of who has produced the report, their competence and who they were contracted by • Any other relevant background information
2.	Monitoring methods	• A description of the methods, timing, level of effort and personnel to be used to undertake the monitoring proposed • This section will be best subdivided by habitat or species that is the subject of the proposed monitoring
3.	Reporting and feedback	• Details of the mechanism and frequency of reporting, and who will receive the monitoring reports
4.	Programme	• A programme of the monitoring proposed, by habitat or species and by year of monitoring
5.	References	• A list of any documents referred to in the text (or these can be provided as footnotes)
	Figures	• Appropriate figures to illustrate the areas for the monitoring
	Appendices	• Detailed survey methods (if appropriate) • Photographs (if appropriate)

Appendix G: Suggested headings and sub-headings for ecological method statements

(excluding those produced to accompany licence applications, the structure and contents of which will be prescribed by the relevant licensing authority)

Based on the suggested structure and contents of Precautionary Working Method Statements set out in CIEEM's *Guidance on Ecological Survey and Assessment in the UK During the Covid-19 Outbreak* (published in 2020), amended for circumstances beyond those covered by the Covid-19 outbreak.[1]

I would like to thank CIEEM for allowing me to reproduce much of Appendix 11 of that CIEEM guidance document, which provides a minimum content of a 'Precautionary Working Method Statement', and which I have adapted here. I was involved in its original drafting, along with other members of the team responsible for producing the overall CIEEM document, particularly Mike Oxford CEcol FCIEEM (representing ALGE), Paola Reason CEcol CEnv MCIEEM (RSK Biocensus), Sue Hooton CEnv MCIEEM (representing ALGE), Martina Girvan CEcol MCIEEM (Arcadis) and Bob Edmonds CEnv MCIEEM (SLR Consulting).

---

1 CIEEM (2020). *Guidance on Ecological Survey and Assessment in the UK During the Covid-19 Outbreak*. Version 3. Published 29 June 2020. Chartered Institute of Ecology and Environmental Management, Winchester, UK.

Heading/sub-heading		Content
	Title Page	
	Contents Page	
1	Background information	• Site name • Site address • Ordnance Survey Grid Reference (or equivalent) • Site location map (with a suitably scaled Ordnance Survey base) • Name and contact details of developer • Name and contact details of contractors involved with the works (as far as they are known) • Name, contact details and evidence of the competence of the ecologist that has produced the method statement • Description of the proposed works • Description of the purpose of the works • Planning status (including reference numbers) – does the project have consent? Is it a permitted development (or equivalent)? Is this method statement accompanying a planning application or discharging a planning condition? • Legislation relating to the species concerned • Justification for a licence not being required (where works affecting the species in question could be licensed) • Consideration of other environmental constraints • Reference to guidance documents used to inform the preparation of the method statement
2	Site information and survey	• Description of site location and habitats (including surrounding habitats) • Description of desk study undertaken, including sources used and dates of searches • Description of field surveys undertaken, including details of the methods used, competence of personnel involved, level of effort, dates, times, weather conditions, etc. • Survey results summary, to include: status of the population (assessed in a national, regional and local context if appropriate), assessment of habitat quality, estimate of population size to be affected, etc. • Detailed results to be provided in an appendix if appropriate • Clear identification of whether the survey accords or does not accord with current good practice guidance (and justification and explanation of the implications if it does not accord) • Summary of any survey limitations and an explanation of the implications of these • Appropriately scaled map(s) showing survey area and results • Photographs of site and specific habitat features, as necessary

Heading/sub-heading		Content
3	Impact assessment	• Quantity (in hectares/m^2 or linear metres, as appropriate) and type of habitat permanently lost in relation to the species concerned • Quantity (in hectares/m^2 or linear metres, as appropriate) and type of habitat temporarily lost in relation to the species concerned, and specify the time frame of the loss • Quantity (in hectares/m^2 or linear metres, as appropriate) and type of habitat permanently damaged in relation to the species concerned • Quantity (in hectares/m^2 or linear metres, as appropriate) and type of habitat temporarily damaged in relation to the species concerned, and specify the time frame of the damage • Identification and assessment of other impacts on the species concerned, such as fragmentation, risk of killing or injury, pollution, disturbance, increased predation • Assessment of the overall impact of the works proposed on the population of the species concerned, with reference to appropriate contextual information • Scaled map to show impacts
4	Mitigation and compensation strategy	Describe the measures to be employed to avoid/minimise impacts, including, for each measure: • Justification for the measure to be used – is it a recommended measure in relevant good practice guidance, or not? If not, why is it proposed? • Likely effectiveness of measure with justification, based on good practice guidelines and/or relevant research • Quantity (in hectares/m^2 or linear metres, as appropriate) of any new habitat being created, or existing habitat being improved • Full details of any capture methods, including timings • Full details of any receptor areas being used to release animals into, including location, details of existing populations, habitat links, evidence that they are within the local range of the species, site ownership • Design drawings of specific features, such as reptile hibernation sites, bat boxes, and details of materials to be used • Details of persons and their roles and responsibilities for implementing the mitigation/compensation works • Details of any operations needing to be overseen by an ecologist • Details of any toolbox talks or signage required to raise awareness and ensure appropriate behaviours • Name, contact details and requirements for the competence level of ecologists overseeing any specific operations • Details of specific machinery or equipment to be used • Details of how any wastes arising from mitigation/compensation works will be disposed of • Scaled map(s) to show extent and location of mitigation/compensation measures
5	Emergency provisions	• What should happen and who needs to be contacted/informed when the provisions of the Method Statement are not followed and/or species are found in unexpected circumstances

// ctd.

Heading/sub-heading		Content
6	Monitoring	• Proposals for monitoring, including methods, timing, survey effort, personnel competence level, frequency, start and end dates • Details of how monitoring will be reported and to whom • Details of baseline to be used and criteria for determining success/failure • Mechanisms for remediation
7	Management	• Details of responsibility for any ongoing management or maintenance of habitat/features
8	Timetable	• Start and finish dates for all activities proposed, identifying activities that are seasonally constrained (i.e. must take place at a specific time of year) and any assumptions made with dates that may change, such as start of construction or phases of development
9	Declaration	• A form to be provided for site operatives to sign and date to confirm that they have read and understood the Method Statement and will implement it
10	References	• As appropriate
11	Supporting figures	• As needed
12	Supporting appendices	• As needed

Index

Abbreviations 40, 41
Access
  to resources 58, 59
  to a site 79, 80, 115
Acronyms 40, 41
Active voice 28
Adjectives 30
Appendix/appendices 31, 51, 56, 57, 68, 69, 72, 81, 82, 83, 84, 86, 107, 111, 112, 113, 117, 125, 126, 127, 128, 129, 134, 138, 141, 142, 158, 162, 163, 175, 177
Association of Local Government Ecologists (ALGE) 154, 166, 192
Audience 4, 8, 9, 12, 13, 15, 16, 17, 18, 20, 21, 55, 61, 73, 90, 91, 98, 101, 104, 107, 108, 111, 113, 114, 117, 118, 119, 121, 125, 128, 134, 138, 139, 143, 144, 145, 146, 147, 148, 149, 150, 151, 155, 167, 168, 175, 178, 184
Autonomy 26

Balanced 8, 90, 98, 100, 101, 103, 132, 145, 177, 188
Baseline conditions 36, 46, 57, 83, 152
Bias/biased 8, 90, 102
Bibliography 186
Biodiversity Action Plan (BAP) 16, 38, 62, 180
Biodiversity benefit 57
Biodiversity Net Gain 18, 57, 108
Biodiversity Net Gain Report 108
Biodiversity Strategy 108
Brackets 40, 41, 106, 171, 181, 182
Breaks 61, 163
Breeding Bird Survey (BBS) 77
British Standard 6, 11, 18, 37, 57, 81, 106, 191
British Trust for Ornithology (BTO) 77
Bullet point(s) 38, 39, 40, 70, 140, 141, 160, 163

Certainty 10, 37, 99, 153, 154, 188
Chartered Institute of Ecology and Environmental Management (CIEEM) 2, 3, 12, 15, 53, 55, 74, 100, 147, 154, 157, 166, 183, 189, 191, 192
Clarity (clear/clearly/clearer) 1, 8, 9, 10, 11, 13, 16, 21, 27, 32, 33, 34, 36, 41, 43, 46, 47, 50, 51, 52, 53, 54, 59, 60, 61, 65, 68, 70, 71, 72, 74, 75, 77, 78, 79, 84, 85, 88, 90, 91, 92, 93, 96, 97, 99, 100, 101, 104, 105, 106, 108, 109, 117, 119, 120, 122, 123, 124, 126, 130, 131, 133, 135, 136, 139, 140, 141, 142, 146, 149, 154, 155, 157, 163, 166, 167, 168, 169, 175, 176, 177, 181, 189, 190, 192
Client 4, 7, 8, 15, 16, 17, 18, 21, 22, 42, 46, 47, 58, 62, 66, 67, 70, 90, 91, 92, 104, 107, 112, 115, 142, 148, 157, 161, 166, 167, 170, 171, 173, 178
Code of Professional Conduct 100
Colon. *See under* Punctuation
Commitment 109, 114, 115, 152

Compensation 7, 15, 36, 57, 106, 116, 150, 154, 177, 181
Competence, competent 2, 3, 8, 14, 17, 19–26, 49, 50, 71, 78, 79, 91, 142, 156, 166, 174, 175, 191
Complaint 3
Concise 8, 9, 10, 47, 90, 104, 109, 113
Conclusion(s) 8, 9, 13, 44, 46, 56, 57, 59, 66, 68, 73, 80, 89, 90, 92, 99, 101, 104, 105, 108, 143, 144, 145, 153, 168, 174, 177
Confidential/confidentiality 85, 123, 124, 133, 168
Conservation Evidence 181
Constraint
seasonal 168
to a development 7, 15, 17, 37, 81, 105, 114, 147, 149, 177
Construction Environmental Management Plan (CEMP) 12, 17
Consultation 18, 34, 63, 141, 159, 160
Consultee 15, 16, 17, 18, 73, 107, 150, 159
Contents page 56, 65, 68, 69, 112, 117, 118, 136, 174
Context 3, 26, 30, 31, 33, 42, 44, 47, 49, 56, 61, 71, 80, 84, 85, 86, 88, 93, 100, 104, 107, 113, 118, 120, 125, 140, 179, 183, 189
Contextual 7, 76, 86, 96, 180, 188
Copyright 67, 82, 123, 193
Cover page 65, 66, 68
Critical review. *See under* Review
Cross-references 51, 81, 119, 128, 157, 158–9, 163, 175, 177
Cumulative effect 57

Data 8, 10, 13, 18, 34, 35, 47, 49, 50, 51, 52, 63, 73, 74, 75, 76, 78, 79, 80, 82, 83, 84, 85, 86, 87, 90, 91, 113, 117, 118, 123, 127, 144, 155, 162, 174, 175, 176, 187, 188, 189, 190
Date
of a desk study record 85, 86
of a report 44, 67, 167
of survey or search 10, 31, 47, 49, 63, 76, 78, 81, 113, 118, 142, 176, 184
Design
of a development 12, 15, 18, 39, 81, 92, 98, 99, 105, 106, 114, 116, 119, 123, 147, 148, 149, 151, 152, 153, 158
of a survey 48, 73, 77, 122, 128, 180
Designated (nature conservation) site 34, 36, 46, 75, 76, 83, 84, 85, 96, 97, 106, 110, 121, 140, 141, 152, 158, 168, 180, 184
Desk study 34, 41, 47, 48, 50, 52, 56, 69, 73, 74, 75, 76, 79, 83, 84, 85, 94, 115, 140, 141, 142, 152, 153, 168, 175, 180, 188, 189
Disclaimer 79
Discussion 10, 11, 13, 31, 34, 45, 46, 56, 57, 78, 87, 89, 109, 110, 113, 114, 176
Double negative 43
Drawing 31, 62, 86, 118, 119, 127, 156, 175

EcIA Checklist 154, 166, 167, 178, 192
Ecological Appraisal 16
Ecological Impact Assessment (EcIA) 12, 74, 81, 96, 97, 115, 119, 147, 152, 157, 158, 189, 192
Ecological Impact Assessment (EcIA) Report 1, 5, 12, 15, 16, 22, 23, 36, 37, 46, 55, 57, 63, 67, 73, 81, 83, 84, 89, 96, 97, 98, 99, 103, 106, 108, 110, 113, 119, 128, 134, 135, 140, 142, 144, 146, 148, 149–50, 151, 152, 153, 154, 155, 156, 158, 159, 160, 167, 178, 180, 181, 189, 191, 192
Editing 64, 72, 131, 136

EIA Co-ordinator 156, 157
Emotive language 30, 104
Endnotes 185
Enhancement 7, 15, 57, 92, 119, 147, 150, 154
Environmental (Impact) Statement 6, 15, 22, 155, 156, 157, 158, 159, 160
Environmental Impact Assessment (EIA) 6, 22, 23, 155, 156, 157, 159
Evaluation 56
Evidence 8, 10, 11, 13, 18, 29, 47, 49, 50, 53, 54, 79, 87, 90, 91, 93, 95, 96, 97, 98, 99, 102, 126, 179, 181

Fact/factual/factually 8, 10, 13, 18, 20, 25, 30, 32, 45, 46, 47, 48, 49, 50, 51, 52, 53, 54, 72, 80, 87, 90, 92, 93, 103, 111, 113, 124, 128, 138, 142, 156, 162, 176
Field survey 10, 41, 42, 47, 48, 58, 59, 63, 73, 74, 75, 77, 79, 81, 82, 84, 86, 93, 115, 141, 142, 168, 175
Figure(s) 30, 31, 42, 43, 51, 68, 69, 73, 78, 86, 87, 106, 117, 118, 119–24, 141, 156, 162, 163, 168, 175, 177
First impression 65, 68, 118
Font, size, type 43, 130, 135, 163
Footnotes 185, 186
Future proof/proofing 188, 189, 190

Good practice guidelines/guidance 3, 25, 48, 49, 52, 53, 54, 57, 77, 80, 82, 91, 102, 113, 142, 152, 168, 175, 176, 177, 179, 181, 193

Habitats Regulations Assessment 17, 108, 193
Harvard system 183
Headers and footers 44, 133, 136
Headings/sub-headings/sub-sub-headings 8, 39, 41, 42, 44, 55, 56, 57, 68, 69, 73, 74, 77, 79, 80, 84, 89, 90, 91, 92, 105, 118, 130, 132, 134, 135, 136, 158, 163, 174
Highlight(s)/highlighting 11, 21, 35, 39, 44, 63, 67, 83, 98, 99, 100, 103, 109, 114, 139, 141, 148, 157, 192

Illustration 127
Impact assessment 89, 100, 153, 176. *See also* Ecological Impact Assessment; Environmental Impact Assessment
Impartial/impartiality 8, 30, 47, 90, 100, 101, 102, 103, 104, 148, 177
Impersonal style 28, 29
In Practice 11, 67, 94, 181, 184, 192, 193
Incidental records/incidentally recorded 85, 88, 124
Interpretation 8, 13, 30, 47, 48, 72, 80, 89, 90, 91, 92, 93, 104, 105, 107, 114, 123, 128, 176
Introduction 8, 13, 35, 41, 44, 56, 61, 62, 63, 65, 68, 70, 71, 91, 110, 111, 135, 142, 151, 154, 174, 180
Irrelevant 47, 51, 87, 133, 136, 137, 165
Italics 41, 44, 171, 182, 183, 184

Joining words 33
Joint Nature Conservation Committee (JNCC) 77
Justify/justified/justification/justifiable 8, 13, 47, 48, 68, 74, 90, 93, 96, 97, 98, 115, 134, 168, 174, 176, 177, 179, 180, 181, 190

Key characteristics 8, 9, 13, 14, 22, 26, 30, 44, 47, 53, 61, 72, 90, 91, 109, 117, 146, 148, 174
Knowledge 3, 10, 11, 13, 19, 20, 21, 117, 118, 138, 156, 188

Latin name 40, 41, 144, 171

Layout (of a report) 66, 130, 131, 133, 158, 159
Legislation 3, 15, 17, 23, 30, 44, 46, 50, 62, 105, 106, 110, 111, 112, 132, 137, 150, 151
Licence(s)/licensed/licensing 15, 16, 17, 50, 55, 78, 116, 123, 142, 147, 149, 151, 172, 175
Limitation(s) 8, 46, 59, 74, 78, 79, 80, 81, 82, 90, 110, 113, 128, 142, 168, 176, 177, 188, 191
Line spacing 43, 135
Lists. *See* Numbered lists
Local Environmental Records Centre (LERC) 52, 75, 82, 180
Local Planning Authority (LPA) 4, 7, 12, 15, 17, 18, 21, 22, 62, 63, 72, 100, 103, 105, 107, 141, 150, 173, 178

Management Plan (Habitat, Landscape and Ecological) 6, 16, 57, 67, 89, 107, 108, 176
Map 48, 58, 62, 73, 75, 76, 78, 85, 86, 87, 119, 120, 122, 123, 141
Method Statement 17, 55, 57
Methods 6, 8, 13, 15, 16, 35, 41, 42, 46, 47, 48, 49, 56, 57, 59, 61, 63, 68, 69, 71, 72, 73, 74, 75, 77, 78, 79, 80, 81, 82, 89, 91, 107, 112, 113, 118, 120, 121, 123, 128, 135, 136, 143, 152, 156, 157, 160, 168, 175, 176, 180
Misinterpretation/misinterpreted 43, 50, 105, 106, 176
Mislead/misleading 52, 80, 99, 135, 144, 178
Mitigation 5, 7, 15, 16, 17, 29, 36, 37, 46, 49, 52, 53, 54, 57, 89, 92, 97, 99, 102, 114, 115, 119, 127, 143, 147, 149, 150, 152, 153, 154, 158, 159, 168, 176, 177, 181, 183, 193
Mitigation strategy 6, 12, 17, 193
Monitoring report 16, 108, 181
Monitoring strategy 16

Non-target 51, 88
Numbered lists 38, 39

Office 5, 58, 59, 60, 61
Opinion(s) 8, 10, 11, 13, 18, 20, 29, 30, 45, 46, 47, 48, 49, 50, 51, 52, 53, 54, 72, 87, 89, 90, 93, 97, 98, 99, 101, 102, 103, 111, 138, 144, 156, 174, 176, 180, 181, 188, 193
Ordnance Survey 36, 48, 62, 75, 76, 84, 96, 120, 122, 123

Page number(s)/numbering 42, 44, 68, 69, 118, 136, 184
Paragraph (forming, spacing) 34, 43, 135
Passive voice 28, 29
Personal style 28, 29
Photographs/photos 10, 43, 48, 61, 62, 69, 75, 76, 86, 87, 117, 118, 120, 122, 124, 125, 126, 127, 128, 163, 175, 177
Planning application(s) 12, 32, 33, 63, 92, 149, 150, 153, 154, 173, 192
Planning condition(s) 5, 15, 16, 17, 150, 158, 191
Plans 17, 58, 61, 62, 119
Policy 3, 15, 23, 36, 44, 46, 62, 106, 115, 145, 151, 184
Precise 8, 47, 72, 75, 82, 86, 90, 99, 104, 114, 117, 124, 176, 177
Preliminary Ecological Appraisal (PEA) 12, 105, 115, 147, 192
Preliminary Ecological Appraisal (PEA) Report 1, 12, 15, 37, 55, 56, 63, 73, 114, 119, 134, 146
Professional judgement 10, 11, 47, 89, 93, 100, 103, 156, 191, 192, 193
Proof of Evidence 18

Proofread/proofreading 68, 133, 154, 161, 162, 163, 164, 165, 166, 169, 170
Proportionate/proportionality 8, 11, 90, 100, 104, 109, 110, 113, 114, 115, 116, 132, 143, 177, 192
Punctuation 26, 27, 32, 38, 39, 138, 163, 165
 colon 38
 semicolon 38, 39, 182
Purpose/purposeful 6, 8, 9, 12, 13, 15, 16, 17, 18, 55, 56, 61, 62, 66, 70, 71, 80, 88, 89, 90, 91, 92, 104, 108, 110, 111, 112, 117, 121, 124, 132, 142, 144, 145, 146, 147, 149, 150, 151, 155, 166, 167, 168, 169, 174, 175, 176, 177, 178, 179, 180, 182, 192

Qualifications 19, 20, 22, 23, 71, 78
Quality Assurance 68, 154, 161, 169, 170, 174
Quotations 44

Rationale 73, 74, 77, 82
Recommendation(s) 8, 13, 29, 46, 53, 56, 59, 89, 90, 91, 92, 104, 105, 109, 114, 115, 119, 149, 151, 152, 168, 176, 181, 191, 193
Reference number 63, 67, 69, 119, 137, 167
Reference(s)/sources 10, 51, 52, 53, 54, 56, 58, 59, 61, 76, 77, 82, 92, 93, 95, 96, 97, 98, 99, 102, 112, 133, 135, 144, 152, 168, 179, 180, 181, 182, 183, 184, 185, 186, 191
Relevance/relevant 2, 3, 13, 15, 17, 18, 22, 23, 29, 31, 36, 41, 46, 47, 48, 49, 50, 51, 53, 54, 55, 59, 62, 63, 68, 70, 71, 74, 75, 76, 77, 78, 80, 82, 83, 84, 85, 86, 87, 88, 90, 91, 93, 95, 99, 100, 102, 107, 110, 111, 112, 114, 116, 122, 124, 128, 129, 130, 132, 133, 134, 135, 136, 137, 139, 140, 141, 143, 144, 145, 148, 150, 151, 152, 155, 158, 165, 168, 170, 173, 175, 176, 177, 180, 181, 182, 189, 191, 193
Repetition/repetitive 91, 112, 128, 158
Report template 6, 72, 111, 129, 130, 131, 132, 133, 134, 135, 136, 137, 192, 193
Research (including reports, articles, papers) 3, 16, 53, 54, 89, 93, 98, 108, 176, 181
Results 6, 8, 13, 15, 16, 35, 36, 38, 41, 46, 50, 51, 56, 57, 58, 59, 61, 63, 68, 71, 72, 73, 79, 80, 82, 83, 84, 86, 87, 88, 89, 91, 93, 105, 107, 113, 114, 118, 121, 122, 123, 128, 149, 152, 155, 168, 175, 176, 190, 192
Review
 critical 167, 173, 174, 178
 technical 2, 68, 71, 154, 161, 166, 167, 169, 173, 178
Robust 6, 8, 9, 13, 47, 72, 82, 90, 91, 92, 113, 153, 155, 174, 175, 189
Royal Society for the Protection of Birds (RSPB) 40

Scientific name. *See* Latin name
Scope/scoping 20, 27, 62, 66, 73, 74, 77, 107, 152, 153, 159
Semicolon. *See under* Punctuation
Sentences 25, 27, 28, 29, 30, 32, 33, 34, 38, 39, 41, 43, 51, 52, 140, 143, 144, 181, 183
Socio-economic 103
Species names 40
Statutory Nature Conservation Body 15, 17, 18, 73, 116, 150
Structure/structured (of a report) 27, 38, 53, 54, 55, 57, 91, 92, 134, 151, 154, 156, 158, 167, 191. *See also* Well structured

Study area 47, 48, 74, 79, 120, 121, 123, 142, 175
Sub-headings. *See* Headings
Sub-sub-headings. *See* Headings
Summary/summaries 8, 44, 56, 65, 66, 68, 70, 86, 108, 113, 117, 118, 119, 121, 138, 139, 140, 141, 142, 143, 144, 145, 149, 151, 158, 160, 168, 178, 190

Table 10, 43, 49, 51, 69, 81, 84, 86, 87, 88, 113, 117, 118, 119, 127, 138, 158, 162, 163
Targeted 8, 9, 47, 61, 85, 90, 91, 94, 121, 142
Technical review. *See under* Review
Template (report). *See* Report template
Tense(s) 35, 36
Time of day 49
Time frame 133, 189
Title page 56, 63, 65, 66, 67, 68, 130, 135, 167, 170, 174
Transparent 8, 13, 47, 90, 93, 99, 138, 175
Truthful 8, 13, 47, 90, 93, 175

Unambiguous 8, 47, 90
Unbiased 30, 101, 148, 169

Valid/validity 73, 78, 79, 121, 174, 175, 176, 187, 189, 190, 192
Version (number) of a report 44, 67, 137, 167, 170, 171, 174

Weather 10, 47, 49, 63, 73, 79, 80, 81, 93, 113, 118, 142, 172, 176
Web-based sources 75, 76, 181
Well structured 8, 13, 90, 91, 92, 96, 174

Zone of Influence 83, 86, 152

CPSIA information can be obtained
at www.ICGtesting.com
Printed in the USA
BVHW031746211220
596004BV00003B/13

9 781784 272418